Research Methods for Global Marketing Practice

MSc in Global Marketing Practice

Research Methods for Global Marketing Practice

First edition October 2010

ISBN: 9780 7517 8415 2
e-book: 9780 7517 8426 8

British Library Cataloguing-in-Publication Data
A catalogue record for this book has been
applied for from the British Library

Published by

BPP Learning Media Ltd
Aldine House, Aldine Place
London W12 8AA

www.bpp.com/learningmedia

Printed in Great Britain

We are grateful to the GMN Advisory Council
members and Paul Weeks for their feedback.

Author:
Dr Kellie Vincent

Your learning materials, published by
BPP Learning Media Ltd, are printed on paper
sourced from sustainable, managed forests.

A note about copyright

Contents

Research Methods for Global Marketing Practice

Congratulations

on choosing to study towards MSc Global Marketing Practice.

You have made the right decision because you have at your fingertips a wealth of support to help guide you through your studies with BPP Learning Media, Ashcroft International Business School and Global Marketing Network.

GMN's CEO Darrell Kofkin summarises the key value of The Global Marketer Programme as:

> *"The programme is designed to prepare you thoroughly for a rewarding career in marketing and ensure you can make a profitable contribution to business anywhere in the world."*

For the first time, you can achieve an academic award at Masters level from one of the UK's largest and most progressive universities, while at the same time ensuring you meet all the academic criteria required for acceptance as a Professional Member of Global Marketing Network.

It is our intention that at this final level of this innovative Masters programme, MSc Global Marketing Practice, you are able to develop the skills you require to perform effectively from day one.

This Study Text is designed to work in conjunction with the following elements of the overall programme:

- A Student Handbook designed to give you all the information you need to achieve success

- A personal E-Tutor to guide, encourage and develop your thinking and provide guidance for your assignment

- Access to The Global Marketer virtual learning environment providing you with a week-by-week plan of study for each module, with access to articles, case studies, additional learning activities and a global discussion forum connecting you with other programme participants

- Access to the Anglia Ruskin University digital library, providing access to an extensive range of online databases, world leading journals and E-books

- Access to the GMN online network with all the latest news about approved conferences, workshops and events that you can attend to support your continuing development

Each chapter of the Study Text is designed to refer to each Study Session within the programme.

Integrating Global Marketing Network Philosophies

Each module has at least one GMN Advisory Council or Faculty member acting as Module Advisor. Module Advisors contribute to the development of the curriculum and study materials.

This module has received input from a number of advisors including Paul Weeks, Programme Leader, Ashcroft International Business School.

Additional information and reviews of the Study Materials by Module Advisors can be found on the Global Marketer Programme supporting Virtual Learning Environment.

Using this Study Text

This Study Text includes features designed specifically to make learning effective and efficient.

- Each chapter begins with an **introduction** to set the chapter in context.

- Chapter **learning outcomes** summarise what you are expected to be able to achieve through reading the chapter. The content covered is summarised in a Chapter Roundup which is organised according to these learning outcomes at the end of the chapter.

- Harvard referencing is used throughout with **a full reference list** at the end of each chapter.

Throughout the Study Text, there are also special aids to learning. These are indicated by the following symbols.

Definition

Definitions are provided for key terms.

Activity 1

Activities for you to complete are included throughout the chapters. Often these will ask you to relate theories to your own organisation. Where appropriate activity debriefs are included at the end of the chapter.

Global case study

Real life global examples are used to illustrate marketing practice.

 Readily available key text book resources, significant chapters and additional reading are selectively added to the chapters to help develop your further reading.

 Selected online materials which link to the topics covered within the chapter are suggested at appropriate points.

Assessment advice

Points to bear in mind in relation to specific chapter topics when preparing your assignment are inserted at relevant places in the chapter.

GMN viewpoint

GMN experts' opinions, snippets from presentations and papers are included to bring both practitioner and academic cutting edge perspectives into the material.

About this module

This module provides you with the appropriate knowledge, skills and abilities you will need to effectively carry out effective business/management research.

A particular emphasis will be placed upon your development towards your practice-based Masters major project. A focus is given to the specific issues faced by managers when carrying out research in an organisational setting. These will include the philosophical aspects of enquiry in social settings, operating in political contexts, negotiating access to key individuals and data, and meeting the research outcomes expected by multiple organisational stakeholders. Consequently, this module will focus on providing the skills necessary to meet these challenges so that you can effectively plan, carry out and report upon your Masters level major project.

The module content contains:

- The nature of management research and its significance for practice.

- Evaluating and developing theoretical / conceptual frameworks – including literature reviewing.

- Evaluating and developing methodological approaches to research – philosophy, strategy, design.

- Data collection / analysis techniques and tools – quantitative and / or qualitative research.

- Practicalities of research – access, acceptance, relationship management, information transfer.

- Communicating research – writing for different audiences.

Module aims

Both the study mode and module content further facilitate the application of, and reflection on, theory in practice. You will be expected to work on an research proposal related either to your own organisation or a national/international business organisation.

The module supports the development of relevant employability, promotability and professional skills which enable you to develop an understanding of the organisation.

Learning outcomes for the module are to:

1   Demonstrate a critical understanding of the different approaches to research used in business/management.

2   Identify and justify decisions regarding their chosen topic, research questions and research methodology.

3   Synthesise and evaluate the current theoretical and methodological developments in their chosen field of study.

4   Demonstrate the required skills and abilities needed to successfully plan, organise, undertake and communicate the findings of a piece of business / management research.

A note on pronouns

On occasions in this Study Text, 'he' is used for 'he or she', 'him' for 'him or her' and so forth. While we try to avoid this practice it is sometimes necessary for reasons of style. No prejudice or stereotyping according to sex is intended or assumed.

Key contacts

Please email globalmarketer@bpp.com for any queries about this Study Text and The Global Marketer Programme.

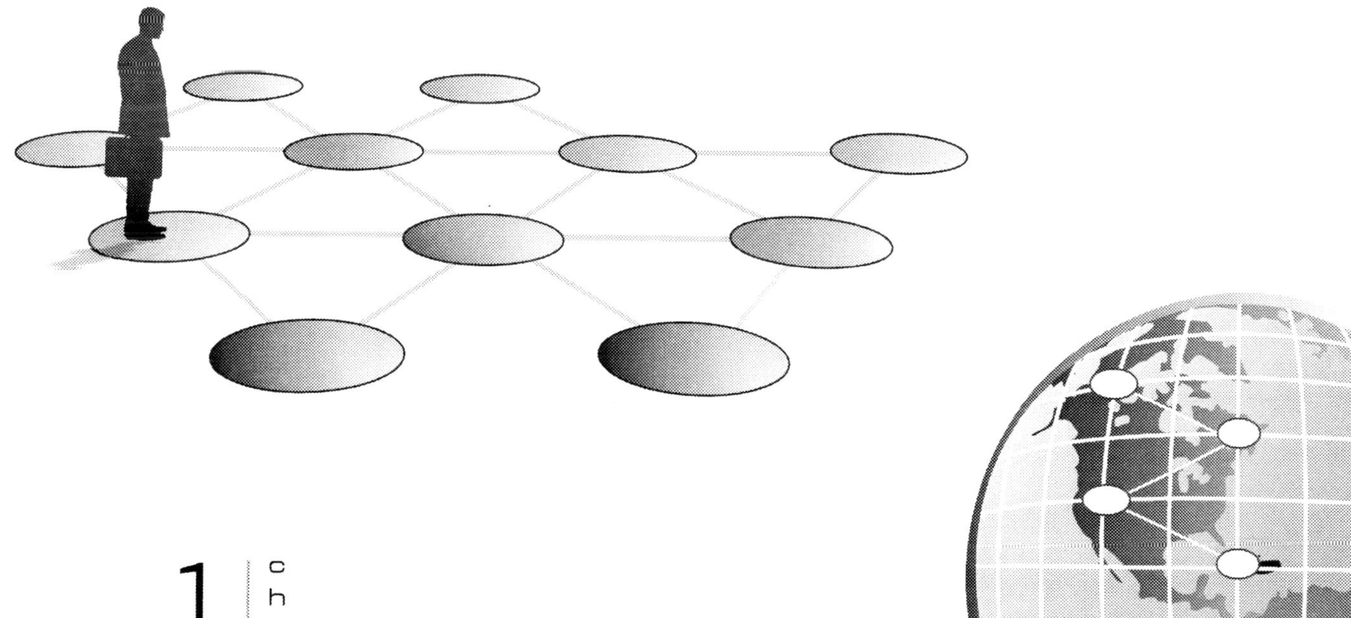

1

The nature, theory and process of academic research

As a marketer you are likely to be making decisions and recommendations which have a significant impact on the performance of an organisation. The quality of the proposed solutions you put forward can be improved if the data you use is systematically gathered and rigorously analysed. The process of completing an academic dissertation or consultancy project will help you to develop your generic management and marketing research skills.

In general terms, research of this nature is most likely to be empirical in nature. It is based on experience and observation. According to Collis and Hussey (2003) it involves the idea of resolutely looking for the required information. In practical terms, management research may often involve a variety of challenges such as gaining access to information and having to deal with the internal politics within an organisation in an ethical manner.

This chapter will introduce the concept of the research project you need to complete for The Global Marketer Programme.

We begin by defining academic research and the research process before taking a detailed look at what distinguishes academic research from consultancy projects.

We also take an extensive look at research strategies and the philosophical view of research you will need to apply.

Contents

By the end of this chapter (in conjunction with the VLE activities) you will be able to:

- Define academic research
- Discuss research philosophies
- Distinguish between different research designs
- Explain the importance of planning to research projects
- Identify the stages in the research process
- Identify the differences and similarities between academic and consultancy style research
- Consider the stages of consultancy investigation
- Review the skills required by researchers when working in a consultancy capacity

1 Definition of research

The research project that you are to undertake to complete The Global Marketer Programme falls into the broader category of business research (you will investigate a marketing related 'business' problem). This fits with Bryman and Bell's (2007) view that there is an inherent practical dimension to business research that should integrate a wide range of disciplines. The authors state:

'business research does not exist in a bubble, hermetically sealed off from the social sciences and the various intellectual allegiances that their practitioners hold' p.5

Definition

Business and management research can be described as a systematic process to find out about business and management (Ghauri and Gronhaug, 2005).

It can also be described as a systematic and organised effort to investigate a specific problem encountered in a work setting, that needs a solution (Sekaran, 2002).

Zikmund (2002) suggests that it is a management tool to reduce uncertainty and involves a systematic and objective process of gathering, recording and analysing data for decision making. The activity must be systematic, not haphazard. It must be objective to avoid the distorting effects of personal bias. According to Ghauri and Gronhaug (2005), business research is based on logical relationships, not just beliefs.

Global case study

The management of a restaurant chain introduces a loyalty card scheme in 200X. In the same year, the company's turnover growth exceeds that achieved the previous year.

The management claim that loyalty cards are an effective way of improving sales growth and believe that their decision to introduce the scheme has been vindicated.

Clearly there has been no systematic approach to confirming the effectiveness of introducing loyalty cards. There has been no systematic consideration of the various factors or variables that might impact on sales levels such as pricing, service style, competition, improvements to store facilities, product choice, product quality, promotion, competition, staff training and the economic climate. Moreover, no attempt has been made to systematically solicit the customers' response to the introduction of the loyalty card and analyse this rigorously. This 'research finding' is therefore not based on logical relationships but purely belief.

Naturally, not all business problems require systematic research. For example, if a company car breaks down in the middle of town, the obvious short-term decision is to have it repaired. However, management research might be required if the company would like to find out the optimal age it should replace its cars from both a financial and operational perspective.

Sekaran (2002) suggests that the difference between a manager who uses common sense alone to analyse and make decisions, and the investigator (researcher) who uses a scientific method, is that a researcher undertakes a systematic inquiry into the matter and proceeds to describe, explain, or predict phenomena based on data carefully collected for the purpose.

1.1 Potential research areas

Assessment advice

Let us consider before going any further the nature of the asssignment you are to complete for this module.

The assignment requires you to produce a **research proposal** for a Masters level dissertation.

Your aim is to identify a suitable topic for the dissertation you are asked to complete for the final MSc. Module research project. Undertaking this task will assist you in understanding the requirements of a postgraduate dissertation and will introduce you to the skills of the literature review and a variety of different approaches to research. It will also encourage you to analyse, evaluate and synthesise information gained from a variety of sources. Completing the work involved with this assignment will help you to complete your final GMP project.

The Task

To submit a **research proposal** for a postgraduate business dissertation that includes:

- A proposed dissertation title.

- An introduction outlining the overall aim, objectives and rationale for your research.

- A critical review of the existing literature in your field of study and how it informs your research.

- A proposed research methodology including a detailed rationale and critical perspective on your:
  - Research strategy and design.
  - Choice of data sources.
  - Choice of data collection methods (tools).
  - Proposed analysis techniques.

- A brief discussion on the ethical implications of your research and how you aim to manage these.

- A useful timetable and plan for the successful completion of the dissertation.

- A reference list and bibliography of the sources used to date, including those giving details of research methodologies.

You can see from the above that **within this particular module you are not expected to complete the Research Project in its entirety but simply to produce a proposal outlining the research you plan to complete**. This module therefore links almost seamlessly into the Research Project for Global Marketing Practice module where you will complete the actual research.

In order for you to prepare your Research Proposal effectively you will be given comprehensive guidance within this Study Text to enable you to actually complete the project (including for example how to complete an academic literature review, plan and use data collection tools and analyse data to establish findings. When you come to undertake the project itself our study materials will not consist of a complete Study Text but supplementary resources including the support of a project supervisor (an E-Tutor). You will also continue to benefit from the support of your cohort peers.

Any review of past dissertation and consultancy topics will reveal a wide diversity in the types of research you could potentially conduct. Potential areas for marketing related management research could be viewed in terms of various stages of the business process as outlined by (Zikmund, 2002):

- Definition of problems facing a business
- Identification of business opportunities
- Clarifying alternatives
- Exploring current programmes and courses of action
- Explaining the outcome of past events
- Forecasting future conditions

The above points are likely to have an important bearing on how you select, plan and manage your research project.

Assessment advice

A topic that is of mutual interest and benefit to both the organisation and yourself should contribute significantly to the success of your dissertation. Consider issues that:

- Interest you

- Are likely to form a major part of the corporate strategy

- You have negotiated to do with your line manager because it was a key element of you joining The Global Marketer Programme.

You must remember overall however that regardless of how strategically important the project is to your organisation it must be something which is achievable within the duration of the programme. You can always design a follow-up programme of research for after you have achieved the qualification!

Recent projects that have been completed at Anglia Ruskin University by students studying for Marketing based Masters awards include the following titles:

- How can accountancy firms formulate new competitive strategies for future prosperity?
- What factors influence consumers' behaviour when purchasing financial products and services?
- How can strategic marketing help developing real estate companies?
- Competitive advantage through knowledge based strategy.
- An evaluation of marketing strategy for X.

Essentially we suggest that you should consider which of the GMP modules you found most interesting and start considering your options from there. We cover how to select your research topic in more detail in Chapter 2.

2 Pure and applied research

In your reading about the nature of management research, you will see references to **pure** and **applied** research. It is important that you understand the difference between them.

Definitions

Pure research, (which is often also called basic, academic or fundamental research) seeks only to understand the process and outcomes of business and management. Sharp *et al* (2003) describe it as the development of theory without any attempt being made to link it to practice. It is highly academic and not at all practical. Its principal aim is the advancement of knowledge rather than to solve a practical problem.

Applied research is directed towards producing results which are of direct relevance to managers and form the basis of recommendations upon which they can act. It will often involve the finding of a solution to a specific problem.

The research you undertake for your research project could be either pure or applied. Research for a consultancy project will most certainly be applied. Sometimes, Masters projects may well include elements of both pure and applied research (as you may have noticed in the examples of projects outlined previously). If your research involves management problem-solving, it will be applied. If your research falls into the empirical investigation category, it will be classified as more pure.

Pure research							Applied research

Pure research

Research was conducted in 2009 by the UK's CBI Higher Education Task Force into the extent that businesses and universities collaborated for mutual benefit. The project assessed the level of current collaboration and identified areas where universities and business should work together in the future. A number of recommendations for a wide range of stakeholders were proposed within the Task Force Report.

You can access the report for yourself at:
http://highereducation.cbi.org.uk/uploaded/CBI_HE_taskforce_report.pdf

Applied research

A chain of roadside restaurants has been experiencing a high level of customer complaints concerning service and food quality. The company has a team of quality inspectors who make secret visits and carry out assessments using a standard score card.

Every outlet is visited at least twice a year and the inspections are taken into account in determining managers' bonuses.

The research undertaken attempts to discover why the inspection linked bonus scheme does not deliver the desired levels of food quality and service satisfaction.

It finds that the company ignored other business success factors such as employee motivation, customer focus, quality driven working practices, teamwork and customer service training.

The above examples demonstrate one of the key differences between **pure research and applied research**. The findings from pure research are usually of importance and value to society in general, whereas the findings of applied research are usually of practical relevance and benefit to managers within organisations (Saunders *et al*, 2009).

3 Research philosophy and theory

3.1 The theory and practice cycle

A perennial question in the field of business and management studies is how theory and practice relate to each other. In the field of research, this link is well acknowledged, and is epitomised by the expression *virtuous circle* of theory and practice (Tranfield and Starkey, 1998). This notion suggests that research is performed on a business and/or management practice and the findings in turn influence the field of theory.

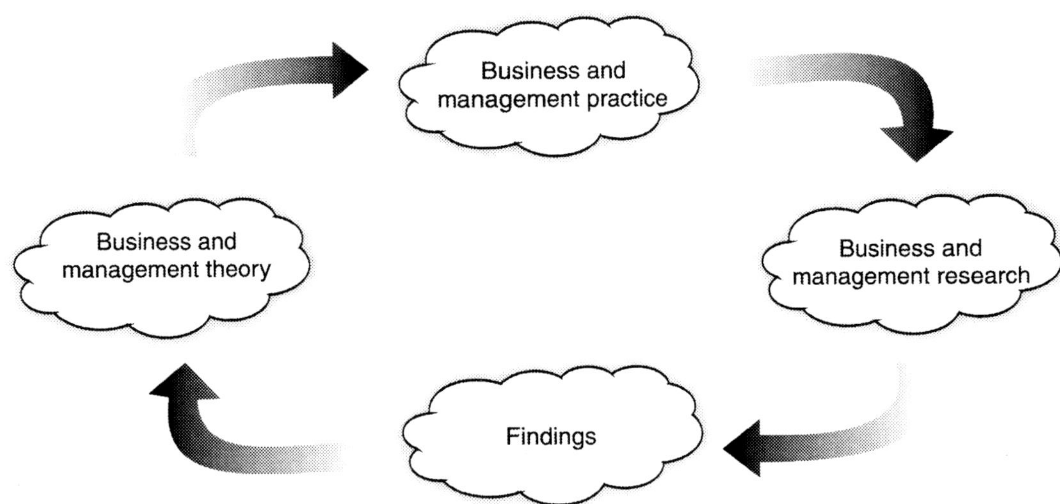

3.1.1 Management perspective

It is perhaps an understandable behaviour for managers to seek to distance themselves from the world of academia and cast themselves in the role of being practical people. As Saunders *et al* (2009) suggest, many managers are dismissive of any talk that smacks of theory, viewing it as something that is all very well to learn at business school but which bears little relation to what goes on in every day organisational life.

3.1.2 Researcher perspective

So from the perspective of your dissertation, it is worth remembering that business and management research has a strong practical orientation sitting alongside an implicit theoretical basis.

There are a number of different ways in which the practical aspect of business and management can inspire the research process, and vice versa.

- Identification and solving of business and management problems
- Providing findings that advance knowledge and understanding
- Applying established theoretical frameworks

The theoretical perspective will be covered in more detail later in this chapter.

3.1.3 Transdisciplinarity

In the earlier modules of The Global Marketer Programme, each topic or subject was studied and examined separately. However, in practice, organisational wide issues and management problems often cross several management disciplines. This is especially relevant in modern business where the organisation is likely to be structured and operates in multi-disciplinary (global) teams or groups rather than being structured along traditional lines in functional departments. Most Masters level research projects are therefore likely to be of a transdisciplinary nature.

A food manufacturer has responded to stock market expectations regarding profitability by gradually increasing its prices and tinkering with its recipes and using lower priced ingredients. However, discerning consumers have noticed the reduction in value for money and consequently sales have fallen. The manufacturer's image has also been damaged by this policy.

The problems and issues are likely to be transdisciplinary. The key business drivers probably need to be examined – is the company customer led or is profitability the key driver? Aspects likely to need review include product development, production efficiency, sourcing of supplies, positioning and promotion and cost control. Perhaps some research needs to be done on accurately defining the manufacturer's value chain.

3.2 Management research and consultancy

Management research and management consultancy are potentially similar as well as being potentially different. We will look at this area later in the chapter but you should note a few points here.

Generally, the first key distinction between academic management research for dissertation purposes and management consultancy for business organisational purposes, is the nature of the brief. The former is carried out primarily for the purpose of academic research towards gaining a university degree, although it may simultaneously be geared to providing a business sponsor with useful findings. On the other hand, management consultancy is primarily an arrangement between a consultant and a client to provide independent advice and assistance.

Generally speaking, management consultancy is likely to be much more involved in implementing recommendations and managing change than would be the case in academic business research. However, you might well have an arrangement to produce a second report for your sponsor, in which you may be required to focus on how any changes highlighted by your research findings would be implemented in practice.

Assessment advice

Despite the need to retain a practical dimension to your project, you should remember that essentially you are creating an academic piece to meet the needs of the University. This does not rule out the use of the project as a consultancy based project but you should bear in mind that the 'deliverables' and language will be different.

The table below outlines the language differences between academic and consultancy projects.

Academic research	Consultancy
Access	Entry
Defining focus	Contracting
• Proposal	• Terms of reference
Methodology	Diagnosis
• Literature review	• Problem solving
• Data collection (Primary and secondary data)	• Data collection (Primary and secondary data)

Academic research	Consultancy
Analysis	Analysis and intervention
• Critical evaluation	• Conclusions and recommendations
• Conclusions	• Readiness for change
	• Assisting change
Thesis/Dissertation	Project Report

3.3 Scientific enquiry and research

The task of science is to invent theories to explain the real world and to test these theories by rational criteria (Robson, 2009). The 'scientific method' refers to a particular set of procedures which if followed give the seal of approval that the result is 'science'. Robson (2009) proposes that this means that the research is carried out:

- **Systematically** – consideration given to what the research is doing and how and why it is being done.

- **Sceptically** – by subjecting ideas, observations and conclusions to scrutiny.

- **Ethically** – following a code of conduct to safeguard those impacted by research.

You may remember from our earlier discussions that business research should also be conducted systematically and in a planned logical manner. As a scientific researcher you will be exposed to many scientific terms and conceptual issues. **Do not be put off if you are not familiar with these terms.** The first term we will use is **paradigm**.

Definition

Paradigm refers to the progress of scientific practice based on people's philosophies and assumptions about the world and the nature of knowledge; in this context, about how research should be conducted (Collis and Hussey, 2003).

The term paradigm can be used at three different levels:

1   **Philosophical level** – reflecting basic beliefs about the world.

2   **Social level** – providing guidelines on how the researcher should conduct their endeavours.

3   **Technical level** – specifying methods and techniques adopted when conducting research.

Another scientific term you will encounter is **epistemology**.

Definition

Collis and Hussey (2003) and Bryman and Bell (2007) explain that **epistemology** is concerned with the study of knowledge and what we accept as knowledge. Jankowicz (2004) suggests that it is to do with your personal theory of knowing. He argues that this effectively renders down to two matters.

1   What counts as knowledge
2   Once you have decided on the above, what counts as proof or evidence of knowledge

You might consider the idea that if a business cuts its administrative overheads it will increase its profits. You find it easy to accept that this is something worthy of research.

However, recruitment consultants may advise prospective candidates that employers tend to reject men wearing brown or green suits or women with red nail varnish or large earrings. Whether you believe these views, feelings and perhaps even prejudices of people to be a valid area of knowledge will depend on your epistemological outlook. Your epistemological outlook, in turn, will drive your search for a research topic.

If you are being sponsored by an organisation you may need to be aware of its epistemological outlook. If the organisation has a 'hard-nosed' remuneration driven approach to staff management, you are unlikely to make much progress if you would like to research 'touchy-feely' issues such as staff development or motivation.

It is therefore important to be aware of the alternative scientific paradigms that exist. There are two main research paradigms (sometimes also referred to as research philosophies) but they should be regarded as two extremes of a continuum:

- **Positivist** – often considered to be quantitative in nature.

- **Phenomenological** – often regarded as qualitative in nature.

3.3.1 Positivism

If you have a positivistic epistemological or philosophical outlook to research, you are likely to believe that only phenomena which are observable and measurable can be validly regarded as knowledge (Collis and Hussey, 2003).

The key idea of positivism is that the social world exists *externally,* and that its properties should be measured through *objective methods,* rather than being inferred subjectively through sensation, reflection or intuition (Easterby-Smith *et al*, 2002).

The key characteristics of the positivist philosophy is that the *observer* is *independent* of what is being observed. You would maintain some distance from the people involved in your research.

Remenyi (1998) observes that under positivism, the research is independent of and neither affects nor is affected by the subject of the research. It is assumed that there are independent causes that lead to the observed effects, that evidence is critical, parsimony is important and that it should be possible to generalise or to model, especially in a mathematical sense, the observed phenomena.

Global case study

A research study attempts to look at the impact of only two variables, employee remuneration levels and supervision, on employee performance.

Positivism is said to lead to a reductionist approach to research, ie variables are reduced to a bare minimum. Parsimony therefore attracts various criticisms.

- It is impossible to treat people as being separate from their social contexts and they cannot be understood without examining the perceptions they have of their own activities.

- A highly structured *research design* imposes certain constraints on the results and may ignore more relevant and interesting findings.

- Researchers are not objective, but part of what they observe. They bring their own interests and values to the research.

- Capturing complex phenomena in a single measure is, at best, misleading. For example, is it possible to assign a numerical value to a person's intelligence?

3.3.2 Phenomenology

If you are disposed to a phenomenological epistemology or philosophy, you are likely to believe that the world and reality are not objective and external, but they are socially constructed and given meaning by people based on their personal experience (Husserl, 1931).

Phenomenology was founded by philosopher Edmund Husserl when he introduced the term in his book *Ideas: General Introduction to Pure Phenomenology* (1931). Phenomenology describes the structures of experience as they present themselves to consciousness, without recourse to theory, deduction or assumptions from other disciplines such as natural sciences.

Phenomenology places considerable regard on the subjective state of the individual. This qualitative approach stresses the subjective respects of human activity by focusing on the meaning, rather than measurement, of social phenomena (Collis and Hussey, 2003).

The phenomenologist believes the world can be modelled, but not necessarily in a mathematical sense. A verbal, diagrammatic model could be acceptable (Remenyi *et al*, 1998).

Global case study

You are considering the factors that influence the performance of your department.

Using a positivist research approach you develop a theory and hypothesis that link performance to variables such as levels of remuneration and supervision and you might develop a mathematically oriented statement along the lines that performance is a function of remuneration and supervision levels.

Eg, Performance = a.remuneration + b.supervision

However, if you were adopting a phenomenological perspective, you would look at the problem more subjectively, being less structured and perhaps more descriptive. For instance, you might adopt a jigsaw puzzle type of model, where performance was the overall picture and all the motivational factors were pieces of the jigsaw. Your approach would be holistic rather than parsimonious/reductionist.

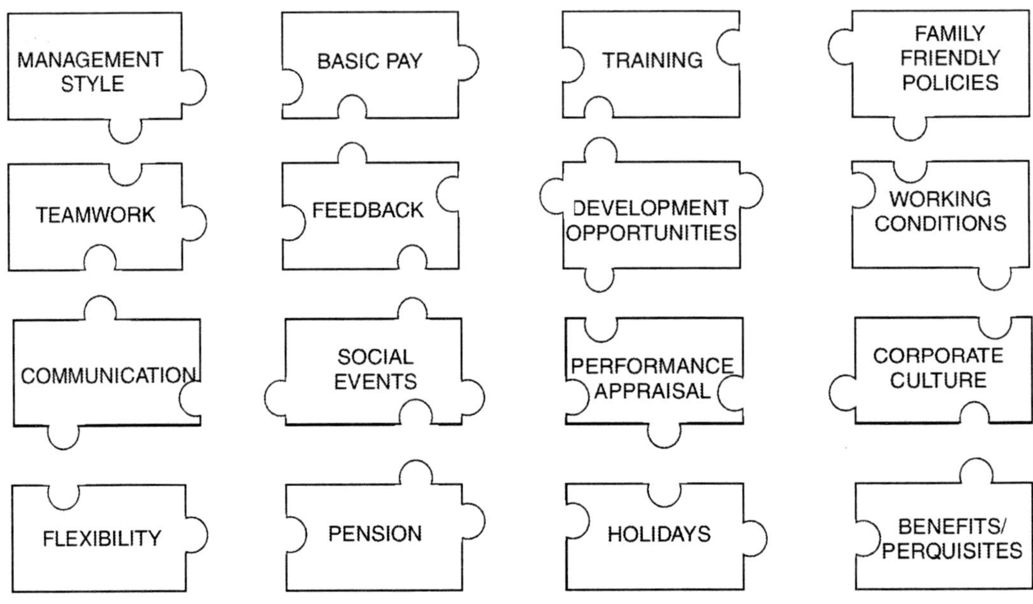

Phenomenological research is sometimes described as the descriptive/interpretative approach and implies that every event is a unique incident in its own right. The approach to phenomenology unfolds as the research proceeds. Early evidence (data) collection suggests how to proceed to the subsequent phase of evidence collection, as does the interpretation of the data itself (Remenyi *et al*, 1998).

Some would argue that it is only phenomenological research that is able to cope with the complexities of business and management. The researcher has to look beyond the details of the situation to understand the reality of the research context (Remenyi *et al*, 1998).

Saunders *et al* (2009) suggest that organisational culture provides an illustration of phenomenological research and refer to Schein's model (as cited in Saunders *et al*, 2009) which talks of organisational culture operating at three levels.

- Visible symbols, eg open plan offices

- Values as revealed in mission statements (if any)

- Underlying assumptions, of which group members are often unaware, but which usually strongly influence how they perceive, think and feel about issues

Saunders *et al* (2009) suggest that the phenomenologist would argue that only research methods rooted in the philosophy of phenomenology offer the opportunity of discovering this vital third level: *'the reality working behind the reality'*.

3.3.3 Post-positivism

Post-positivism attempts to address the criticisms made of positivism. Take for example the notion of the independence of researcher and research subject. Post-positivism accepts that the theories, hypotheses, background knowledge and values of the researcher can influence what is observed (Reinhart and Rallis, 1994). However, there is still a commitment to objectivity, which is addressed by taking into account any bias on the part of the researcher (Robson, 2009).

3.3.4 Pragmatism

There is an ongoing debate between the supporters of positivism and phenomenology as to which approach is 'better', the so-called 'paradigm wars'.

As defined earlier, a paradigm is a philosophical and theoretical framework of a scientific school or discipline within which theories, laws and generalisations, and the experiments performed in support of them, are formulated.

Saunders *et al* (2009) suggest that the rivalry between supporters of the two approaches misses the point as each philosophy is better at doing different things. As always, which is 'better' depends on the research question(s) you are seeking to answer. Business and management research is often a mixture between the two, ie what in some books is referred to as a pragmatic approach.

Pragmatism is a philosophical position with a respectable, mainly American, history (Robson, 2009). The characteristic idea of philosophical pragmatism is that of efficiency in practical application – the issue of 'which works out most effectively' is of primary importance.

Remenyi *et al* (1998) encourage the researcher to draw on whichever epistemological approach is appropriate and if necessary use a variety to *triangulate* the findings and theories by taking this pragmatic approach.

Definition

Triangulation refers to the researcher using multiple sources of evidence to improve accuracy in research conclusions (Robson, 2009; Yin, 1989).

3.4 Theory generation

If we agree with the premise that scientific theory based on facts is relevant to managerial learning, we need to understand what is meant by theory and what its purpose is. According to Zikmund (2002) theory has two purposes

- Prediction
- Understanding

Hence at its most basic level, theories explain how things function and why certain events happen (Black, 1993). Or, as Zikmund (2002) defines it, theory is *'a coherent set of general proposals used to explain the supposed relationships among certain observed events'*. Theories allow generalisations *beyond* individual facts or situations.

Definitions

According to Bryman and Bell (2007) the term **theory** is used in a variety of ways. The authors state that the most commonly applied meaning of theory is as an explanation of observed regularities. These are also referred to **as middle range theories**.

A **theoretical perspective** on the other hand refers to higher levels of abstraction in relation to research findings. These are also referred to as **grand theories**.

Gill and Johnson (2002) put a slightly different flavour to the definition by suggesting that theory is a formulation of the cause and effect relationship between two or more variables, which may or may not have been bred and tested.

Variables can be classified as qualitative or quantitative.

Definitions

Qualitative variables – tend to be 'wordy' and rich in contextual depth. There is usually a deal of 'quality' in the depth of the context. There is usually less 'volume' of qualitative data due to the amount of time detailed analysis takes.

Quantitative variables – quantifiable numerical data. Usually there is a fair 'quantity' of data.

We will return to the important distinction between qualitative and quantitative variables later in the chapter when we discuss research strategy.

Activity 1

Think about the discussion we have raised regarding the differences between postivitist and phenomelogical perspectives. Which perspective do you think would make use of qualitative variables?

Zikmund (2002) suggests that being able to anticipate future conditions in the environment or within an organisation may prove valuable, prediction alone however cannot satisfy the scientific researcher's goal and so 'understanding' is required. In most situations, of course, prediction and understanding go hand in hand. To predict phenomena, we must have an explanation of why variables behave as they do. Theories provide these explanations.

A supermarket chain finds that it maximises its turnover by using a certain store layout, eg fruit and vegetables at the front of the store and beverages at the back.

The chain runs different size stores in different areas and needs to understand customers' buying behaviour to effectively plan store layouts in the future.

Theories provide a structure or guide to marketing and strategy by providing insights into general rules of behaviour. When different incidents are theoretically compared, the scientific knowledge gained from theory developments can have practical value. A good theory allows us to generalise beyond individual facts so that general patterns may be predicted and understood Zikmund (2002).

3.4.1 Types of theories

You may be finding this discussion of theory a little daunting. So perhaps you may find an attempt to stratify theories helpful.

Creswell (2003) summarised theories into the following.

- **Grand theories**. These are usually found in the natural sciences, eg theory of relatively, theory of evolution or big bang theory.

- **Middle range theories**. These lack the capacity to change the way we think about the world. Nevertheless, they are significant.

- **Substantive theories**. These are 'restricted to a particular setting, group, time or problem'.

At Masters level, the programme expectation, as well as the practical time constraints, mean that your research project will most sensibly be done at the substantive theories level.

You will find that some books use the word *models* as an alternative for theories or, more often, to refer to theories with a narrow focus, or substantive theories (Collis and Hussey, 2003).

3.5 Research strategy

Related to the generation of theory is the research strategy employed to develop or test the theory. Bryman and Bell (2007) use the distinction between qualitative and quantitative research because it is a helpful umbrella for a range of issues concerned with the practice of business research. Qualitative and quantitative research also differs with respect to their epistemological foundations.

The following table outlines the fundamental differences between qualitative and quantitative research strategies in terms of the epistemological orientation and role of theory generation methods employed.

	Quantitative	Qualitative
Detail	Quantification	Emphasis on words and context
Epistemological orientation	Positivist	Phenomological
Role of theory generation	Deductive: testing theory	Inductive: generating theory
Implication for data collection methods	Large sample sizes Tools employed to measure concepts	Small samples investigated in depth taking into account the totality of the situation Methods to encourage development of ideas

The research approaches employed in relation to their role in either testing theory or generating theory are either deductive or inductive as shown in the table above.

3.5.1 Deductive research

Definition

According to Zikmund (2002) **deductive reasoning** is the logical process of deriving a conclusion from a known premise or something known to be true.

Global case study

- We know that all marketers are human beings.
- We know that GMN Honorary President Jag Sheth is a marketer.
- Therefore Jag Sheth is a human being.

According to Collis and Hussey (2003) deductive research is a study in which a conceptual and theoretical structure is developed and tested by empirical observation. Saunders *et al* (2009) and Bryman & Bell (2007) refine this to a process of developing a theory and hypothesis/hypotheses and designing a research strategy to test the hypothesis/hypotheses.

Definition

A **hypothesis** is *'an unproven proposition or supposition that tentatively explains certain factors or phenomena; a proposition that is empirically testable'*. Zikmund (2002).

Bryman and Bell (2007) define a hypothesis as:

'An informed speculation, which is set up to be tested, about the possible relationship between two or more variables' p.728

Collis and Hussey (2003) take this a little further by defining a hypothesis as an idea or proposition which you test using statistical analysis. Verma and Beard (1981) suggest that in many cases, hypotheses are hunches that the researcher has about the existence of relationships between variables.

Global case study

You are interested in finding out whether the earning power of bank personal account managers is related to their physical appearance.

You therefore develop a theory that all other factors being equal, earning power is related to a manager's physical appearance.

You develop the following hypothesis.

- A manager's appearance influences the extent to which a customer likes them.
- Liking for a personal account manager increases a customer's trust.
- Trust for a personal account manager increases a customer's amenability to purchase financial products and services.
- Increased customer purchasing improves the personal account manager's assessed performance and remuneration.

Now 'physical appearance', 'liking', 'trust' and 'amenability to purchase' are abstract concepts which are difficult to test empirically. Hence it is necessary to set up a scale by which these concepts can be measured. The technical term is **operationalisation**. In doing this, we create indicators, or measures, which represent empirically observable instances or occurrences of the concepts under investigation. We overtly link the abstract concept to something that is observable and whose variation is measurable (Gill and Johnson, 2002).

A key consideration in hypothesis testing is to ensure that you are focusing on the variables in the hypothesis, with other variables being controlled. Hence in the above example you would need to ensure that each personal account manager had a similar sample of customers, with similar potential sales opportunities.

It might be helpful to try to summarise the steps that reflect the deductive research approach. The following diagram is adapted from the work of Robson (2009).

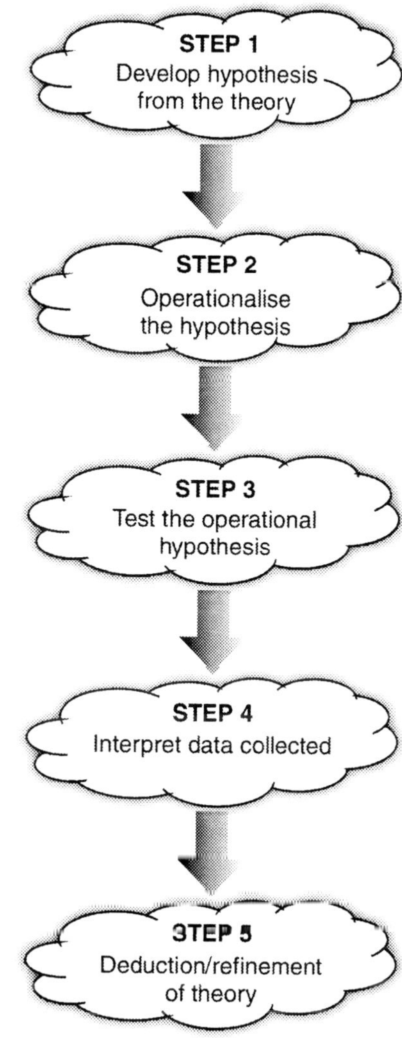

Deductive work approach – adapted from Robson (2009).

3.5.2 Hypothetico-deductive approach

The method of starting with a theoretical framework, formulating hypotheses and logically deducing from the results of the study is known as the hypothesis-deductive method (or hypothetico-deductive) (Sekaran, 2002). This approach is fairly structured unlike the alternative inductive approach (which we will describe later).

When conducting research, sometimes hypotheses that were not originally formulated do get generated through the process of induction. That is, after the data is obtained, some creative insights occur, and based on these, new hypotheses could get generated to be tested later. Generally, in research, hypothesis testing through deductive research and hypothesis generation through induction are both common. The hypothetico-deductive approach therefore makes use of the benefits of both inductive and deductive designs. When there is little existing experience in a particular area it is almost impossible to take a purely deductive approach without making huge assumptions which may be unfounded.

Global case study

A rugby club found that they had no problem in finding players for their junior teams but couldn't fill their adult team.

At the begining of the research project, the hypothesis the manager demanded was integrated into the design was that boys stopped playing rugby at 16 because they were busy socialising with friends and girlfriends and that these demands on their time took precendent.

Despite being a possible explanation worthy of further investigation, nobody on the research team was a 16 year old rugby player and there was no previous research to indicate this was happening in reality. Rather than taking a purely deductive approach and testing this hypothesis alone, preliminary qualitative research was conducted in order to establish what other explanations may exist.

The results from this showed that actually this was not a key variable in the reason why the players left the team but rather that they were used to playing with younger players and not adults. The fear of being badly injured with more experienced (and larger) players was a key concern preventing many from wanting to move on to the adult team.

The project continued with a quantitative survey looking at what factors would encourage players to move to the adult teams including the introduction of a mid level 'colt' bridging team to ease the transition.

Luck (1999) suggests that a hypothesis should be written as a statement that is as simple and unambiguous as you are currently able to make it. It may be a single, short sentence or a group of sentences which make positive general statements such as:

- P is caused by …
- Q depends on …
- R only happens under conditions of …
- The rate of change of S is proportional to …

The hypothetico-deductive approach may be regarded as having various weaknesses.

While its main strength is a clear scientific approach and deduction based on measured results, from a practical standpoint it is not always easy to set up the necessary research.

Global case study

You might like to research whether the bonus scheme introduced by a football club improved its results. You could compare its points tally for the season in which the bonus scheme is introduced against its points total for the previous year when there was no scheme in place. Or you might experiment from match to match, offering win bonuses for some matches but not for others. But you would still face the problem of controlling the other variables, such as quality of the opposition, team selection, refereeing decisions, playing conditions etc which might affect the outcome of matches.

You also need to be aware of a problem that is called *falsification,* an idea first put forward by Popper (1959) when describing his hypothetico-deductive method. This is also known as the method of **'conjectures and refutations'.** The concept demands falsifiable hypotheses, framed in such a manner that the scientific community can prove them false (usually by observation). According to this view, a hypothesis cannot be *'proved',* because there is always the possibility that a future experiment will show that it is false. Hence, failing to falsify a hypothesis does not prove that hypothesis: it remains provisional. However, a hypothesis that has been rigorously tested and not falsified can form a reasonable basis for action, ie we can act as if it is true, until such time as it is falsified. Just because we've never observed rain falling upward, doesn't mean that we never will - however improbable, our theory of gravity may be falsified some day.

Consider the assertion that all swans are white. If following the verification route, the researcher would start travelling around the country accumulating sightings of swans and, provided that he or she did not go near a zoo, a very high number of white sightings would eventually be obtained, and presumably no black sightings. This gives a lot of confidence to the assertion that all swans are white, but still does not conclusively prove the statement.

If on the other hand, the researcher takes a falsification view, he or she would start to search for swans that are *not* white, deliberately looking for contexts and locations where non-white swans might be encountered. Thus, our intrepid researcher would head straight for a zoo, or perhaps book a flight to Australia, where most swans happen to be black.

Hence the initial hypothesis would be falsified, and it might then have to be modified to include the idea that all swans are either white or black.

(Easterby-Smith et al, 2002)

Problems also arise in terms of ensuring not only the validity of the hypotheses postulated, but also their completeness.

3.5.3 Issues with deductivist approaches

A criticism of the deductive approach is its tendency to construct a rigid methodology that does not permit alternative explanations of what is going on. In that sense, there is an air of finality about the choice of theory and definition of hypothesis (Saunders *et al,* 2009).

Deductive research might be conducted to confirm productivity. However, it is unlikely to suggest other reasons for improved productivity such as better staff morale or better leadership.

Alternative theories may be suggested by the deductive approach. However, these would be within the limits set by the highly structured research design (Saunders *et al,* 2009).

Saunders *et al* (2009) suggest that deductive research might be quicker to complete as data collection is based on 'one take'. However time must be allocated to setting up the study prior to data collection and analysis. However, generally the deductive approach can be regarded as low risk and minimises the fear that no real useful data patterns and theory will emerge. They also argue that a topic on which there is a wealth of literature from which you can define a theoretical framework and hypothesis lends itself more readily to a deductive approach.

Saunders *et al* (2009) suggest a final characteristic of the deductive approach is *generalisation.* To actualise this entails selecting sufficiently large samples. By this we mean that the results from the research can be generalised (also relevant) to the wider population.

3.5.4 Inductive reasoning

According to Zikmund (2002) at the empirical level, theory can be developed with inductive reasoning. Inductive reasoning is the logical process of establishing a general proposition on the basis of observation of particular facts. Unlike deductive research where hypotheses are tested, inductive researchers will make qualitative observations in the contextual setting of behaviours before trying to form any predefined notions of what may be occurring.

> **Definition**
>
> **Inductive** – an approach to the relationship between theory and research in which the former is generated out of the latter (Bryman and Bell, 2007).

According to Collis and Hussey (2003) inductive research is the study in which theory is developed from the observation of empirical reality. Thus general inferences are inducted from particular instances, which is the reverse of the deductive method. Since it involves moving from individual observation to statements of general patterns or laws, it is referred to as moving from the specific to the general. Saunders *et al* (2009) summarise this effectively by stating that under the inductive approach, theory would follow data, rather than *vice versa*.

Gill and Johnson (2002) outline the two key arguments justifying the use of an inductive approach in social sciences (and presumably in business and management research).

1   Explanations of social phenomena are relatively worthless unless they are grounded in observation and experience. The nature of deductive theory is speculative and *a priori*. On the other hand, theory that inductively develops out of a system of empirical research is more likely to fit the data and thus more likely to be useful, plausible and accessible.

2   Human action has an internal logic which must be understood in order to make action intelligible, ie it involves an element of subjectivity which the deductive approach does not address. Supporters of the inductive approach to research are concerned with the discovery of this internal logic.

Inductivists reject the deductivists' stimulus-response model of human behaviour as depicted in the diagram below.

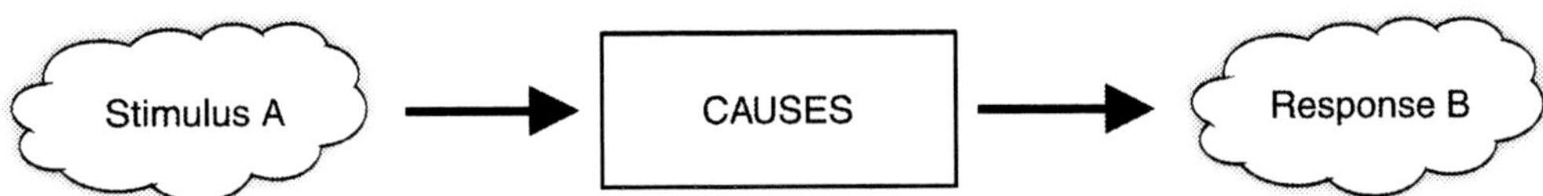

Inductivists instead prefer a model which takes into account the subjectivity of human behaviour.

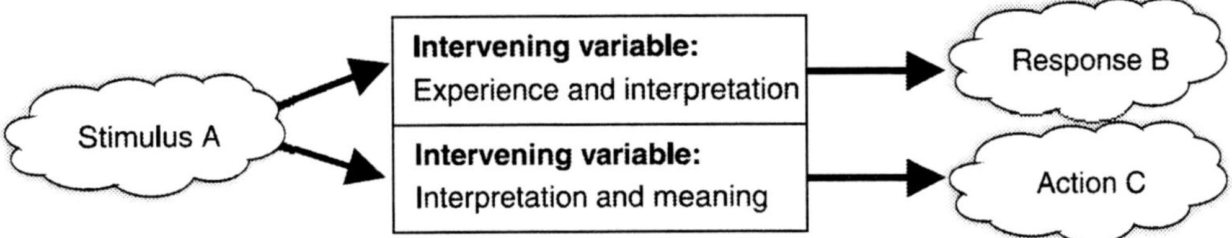

Hence, to achieve the discovery of internal logic, a more unstructured approach is recommended for social science (and management) research that ostensibly allows access to human subjectivity, without creating distortion, in its natural or everyday setting (Gill and Johnson, 2002).

As an aspiring manager or entrepreneur, you should be able to identify with the idea put forward by Wheatley (1992) that we live in a world that is subjective and shaped by our interaction with it.

Refer back to the earlier example relating to the research on the relationship between the earning power of personal account managers and their physical appearance.

If you were using an inductive approach you might begin your research by interviewing a selection of customers for a cross-section of managers based on:

- Sales achieved
- Physical attributes – tall, short, heavy, slim etc

The data gathered from these initial interviews might then help you to focus on key areas where you continue with the next stage of your research. This approach is likely to help you better understand the subjective buying behaviour (internal logic) of the bank's customers.

Hence, with the above approach, you would not begin with a theory and a set of hypotheses, but instead you would begin with data collection and then build up to a theory. In early interviews if you were taking a purely inductive approach you would not even have any predefined questions because you would start from the viewpoint that you do not yet know what it is you need to ask. As the interviews progress learning from previous interviews would mean that more detailed questions could be formed.

According to Saunders *et al* (2009) inductive research can be more protracted than deductive research. Often, the ideas, based on a much longer period of data collection and analysis, have to emerge gradually. However, Saunders *et al* (2009) believe that a less structured approach may reveal alternative explanations of relationships between variables. Research using the inductive approach would be particularly concerned with the context in which such events were taking place. Therefore, the study of a small sample of subjects may be more appropriate than a large number as with the deductive approach.

Saunders *et al* (2009) also suggest that there may be practical reasons for using an inductive approach such as a lack of prior knowledge of the subject and you may therefore simply not be in a position to frame a hypothesis.

4 Research designs

Once an overall philosophy towards the research is adopted, the next stage is to consider the research design to use. There are a number of alternative designs available to the researcher and we will discuss those that are most frequently employed in business research.

4.1 Action research

Collis and Hussey (2003) explain that action research is an approach which assumes that the social world is constantly changing, and the researcher and the research itself are part of this change. The main aim of action research is to enter into a situation, attempt to bring about change and monitor the results.

According to Somekh (1995), action research rejects the concept of a two-stage process in which research is carried out first by researchers and then, in a separate second stage, the knowledge generated from the research is applied by practitioners. Instead, the two processes of research and action are integrated. Elliot (1991) states that in action research, theories are not validated independently and then applied to practice. They are validated through practice alone.

The term action research was originally used by Lewin (1946) who saw the process of enquiry as forming a cycle of

- Planning
- Action
- Observation
- Reflection

(Collis and Hussey, 2003)

Denscombe (2003) has refined this cyclical process as depicted in the following diagram.

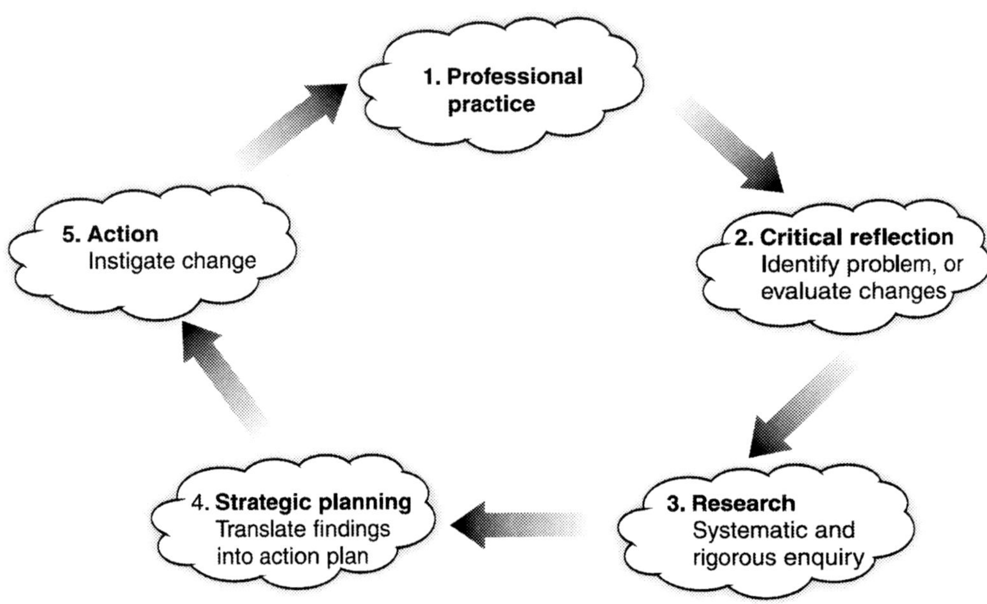

The cyclical process in action research

Saunders *et al* (2009) explains that action research is frequently interpreted by management researchers in a variety of ways, but there are three common themes within the literature.

- Focus on the purpose of the research: management of change (Cunningham, 1995).

- Involvement of practitioners in the research and in particular, a close collaboration between practitioners and researchers, for example academics and external consultants.

- The research should have implications beyond the immediate project; it must be clear that the results could inform their contexts.

The essential practical, problem-solving nature of action research makes this approach attractive to practitioner-researchers who have identified a problem during the course of their work and see the merit of investigating it and, if possible, of improving practice. There is nothing new about practitioners operating as researchers, and the 'teacher as researcher' model has been extensively discussed (Bartholomew, 1971; Cope and Gray, 1979; Raven and Parker, 1981).

Denscombe (2003) summarises the advantages and disadvantages of action research as follows.

4.1.1 Advantages

- It addresses practical problems in a positive way, feeding the results directly back into practice.

- It contributes to the professional self-development of the practitioners involved.

- It should entail a continuous cycle of development and change for the organisation, to the extent that the action research is geared to improving practice and resolving problems.

- It involves participation in the research for practitioners and generally involves greater appreciation of, and respect for, practitioner knowledge.

4.1.2 Disadvantages

- The need for practitioner involvement can sometimes limit the scope and scale of the research.

- The 'work-site' approach affects the representativeness of the findings and the extent to which generalisations can be made on the basis of the results.

- The integration of research with practice limits the feasibility of exercising controls over factors (variables) of relevance to the researcher.

- There may be rivalry between researcher and practitioner over ownership of the research process.

- Action research tends to involve an extra burden of work for the practitioner, particularly at the early stage before any benefits feed back into improved effectiveness.

- The action researcher is unlikely to be detached and impartial in his or her approach to the research.

So action research focuses on the management of change (Cunningham, 1995), involving practitioners and researchers and feeds back into the practitioners' professional practice.

Consultancy also focuses on change: the consultant and client engage to consider what changes are needed. Depending on what consultancy model is used, the consultant may do the research and advise the client. Or they may do it in tandem. In a process consultancy model (discussed later in the chapter), the client would work with the consultant to reflect on what changes are needed, doing the research and translating the findings into action. In this way, they are engaging in action research in their own practice.

Margerison (2001) makes suggestions about managing the relationship between the client and the consultant. There is a suggestion that the consultant uses feedback from data gathered during recorded meetings to reflect back to the client. This can then be used so that they can jointly come to conclusions and decide on any action to be taken.

4.2 Case research

Denscombe (2003) states that case studies focus on one instance (or a few instances) of a particular phenomenon with a view to providing an in-depth account of events, relationships, experiences or processes occurring in that particular instance.

According to Yin (2009), case studies have the following characteristics.

- They are an empirical way of investigating a contemporary phenomenon within its real life context.

- The boundaries between the phenomena and the context are not clearly evident.

- They use multiple-methods of collecting data which may be both quantitative and qualitative.

- They are particularly valuable in answering 'Who?', 'Why?' and 'How?' questions in management research.

Collis and Hussey (2003) state that case studies are often described as *exploratory research,* used in areas where there are few theories or a deficient body of knowledge. However, Scapens (1990) also includes the following.

Type of case study	Overall aim
Descriptive	These are restricted to describing current practice.
Illustrative	The research attempts to illustrate new and innovative practices adopted by particular organisations.
Experimental	The research examines the difficulties in implementing new procedures and techniques in an organisation and evaluating the benefits.
Exploratory	Existing theory is used to understand and explain what is happening.

Case studies may be carried out to follow up and put flesh on the bones of a survey. On the other hand, they can precede a survey and be used as a means of identifying key issues which merit further investigation. However, generally the majority of case studies are carried out on a 'stand-alone' basis (Bell, 2005).

Remenyi *et al* (1998) argues that because of its flexible nature, a case study may be an almost entirely positivistic or almost phenomenological study or anything between these two extremes.

Saunders *et al* (2009) suggests the following benefits of case studies.

- They can be a very worthwhile way of **exploring existing theory**.
- They can enable you to **challenge an existing theory** and also provide a source of new hypotheses, where the case study is well constructed.

The advantages and disadvantages of case studies identified by Denscombe (2003) can be summarised as follows.

Advantages

- Focus on one or a few instances enables the researcher to deal with the subtleties and intricacies of complex social situations. In a large scale survey the processes can be elusive (Bell, 2005).
- The analysis is holistic rather than based on isolated factors.
- It allows for use of a variety of research methods and fosters the use of multiple sources of data.
- There is no pressure to control events.

Disadvantages

- It is vulnerable to criticism in relation to credibility of generalisations. There is a need to demonstrate the extent to which the case is similar, or different from, other cases of its type.
- Susceptibility to accusations of producing *safe data*. Hence careful attention to detail and rigour must be exercised in the use of this approach.
- The *boundaries* of the case can prove difficult to define in an absolute and coherent fashion.
- Negotiating access to case study settings can be a demanding part of the research process.
- The *observer effect* can come into play; those being observed might behave differently from the way they would do usually, because they are being observed.

4.2.1 Different types of case study research design

Yin (2009) describes a research design as a *'logical plan for getting from here to there, where here may be defined as the initial set of questions to be answered, and there is some set of conclusions (answers) about these questions'* p.26. Thus the research design should deliver the answers to the initial questions framed in the case study.

Philliber, Schwab and Samsloss (1980) cited in Yin (2009) state that a research design should address four problems:

- What questions to study
- What data is relevant
- What data to collect
- How to analyse the results

It is vital to develop an appropriate research design so that the data collected suits the original questions asked. Otherwise the data cannot be used to answer the questions set. It is important to ask the right questions and collect the appropriate data to answer those questions.

There are five elements of a research design (Yin, 2009). These are

- A study's questions
- Any propositions
- Units of analysis
- Logic linking data to propositions
- Criteria for interpreting the findings

We look at each of these briefly before we consider single case study and multiple case study design.

- **Study questions**

    Yin (2009) notes that case studies are best suited to 'how' and 'why' questions which are qualitative and often need some detailed analysis. Devising the case study is best tackled in three stages.

    First, read around the subject and use this to narrow the topic(s) of study.

    Second, look closely at key studies which may throw up some leads for further study and questions that should be asked in the research design.

    Third, look at even more studies to support the questions identified already and whether these can be refined.

- **Propositions for study**

    Yin (2009) suggested that propositions frame the purpose for undertaking a case study by expressing opinions that then require proof. He illustrates this by stating a proposition that organisations collaborate to obtain mutual benefits. This proposition can then be tested by asking a question for study which is how and why organisations collaborate to provide joint services. Once the proposition has been stated the researcher can then start to collect evidence to define and ascertain the benefits stated.

- **Unit of analysis**

    This is what is being studied. Cases may be single individuals, groups, events and entities (Yin, 2009). It is highly important to be clear on what is being studied.

    The case must be scoped to include who or what is to be studied and the time line for the study. As the case study develops, you may need to revisit what is being studied in the light of new data and discoveries. Yin (2009) stated that case studies should consider real life examples rather than abstract ideas.

- **Logic linking data to propositions**

    You then need to look at how you will collect data and analyse it so that you can answer the questions set and the propositions stated. If you are planning a lengthy longitudinal study you need to consider time-series analysis and thus ensure you collect time markers as part of your data.

- **Criteria for interpreting the findings**

    Statistical analyses may be used to interpret findings especially when looking at statistical significance but other ways of interpreting findings must be used where statistics aren't needed. Yin (2009) suggests trying to identify and address explanations for your findings other than those expected. If you can do this at the data collection stage you can then prove or disprove data validity when you come to analyse and interpret data.

4.2.2 Single case study

A single case study is one undertaken only once. Yin (2009) gives five circumstances where using a single case study is the best approach. The discussion here looks at critical case studies and representative case studies within the context of single case studies.

- **Representing the critical case**

    This is where a well-formulated theory is being tested. The example cited by Yin (2009) is a Cuban Missile Crisis of 1962 case study published in 1971. The study used three theories to explain the crisis and tested each by comparing them with what actually happened.

- **Representing an extreme or unique case**

  Yin (2009) gives the example of medical conditions where an extreme or unique case is worth documenting and analysing because it is so unusual.

- **Representing a typical or representative case**

  The reason for taking a single case study which is typical or representative is to '*capture the circumstances and conditions of an everyday or commonplace situation*' (Yin, 2009). The intention is to study a typical industry, school and so on to obtain information on the average rather than the extreme or unique case.

- **The revelatory case**

  The single case study approach delves into areas previously unexplored (Yin,2009). He cites the example of a study of unemployed men in the USA where the researcher spent much time with the men. The insights and observations there from led to a significant and famous case study.

- **The longitudinal case**

  In a longitudinal case study, a single case would be studied and revisited when it is anticipated that changes would occur.

4.2.3 Multiple-case study

A study may include several individual case studies and in this way becomes a multiple-case study. These are often used to replicate findings from a single case study. The later experiment is used to duplicate the earlier experiment and thereby prove its findings. In other approaches some of the original conditions considered unimportant, could be varied to see if the original findings remain.

Thus the multiple cases may be used to prove or disprove the original findings and develop a theoretical framework. Each case must be selected so that it predicts similar results (literal replication) or contrasting results, for anticipated reasons (theoretical replication). During the case study process, any findings that alter the original case study design should be fed back into the design. This process can involve selecting other cases or changing the case study (data collection). The number of cases chosen depends on how many replications are required in the study.

Yin (2009) suggests anything from two to three literal replications for a straightforward case to several for a subtle theory where a high degree of certainty is demanded. You also need to think about how many alternative explanations may exist and how important these are.

4.2.4 Validity and case study research

Analytic generalisation

Yin (2009) describes analytic generalisation as way of approaching the generalisation of the case study results. How this works is that an already developed theory is used as a template to compare the empirical results of the case study. If two or more cases are shown to support the same theory, replication may be claimed. The empirical results may be considered more credible where two or more cases support a theory over another rival theory.

Triangulation using multiple methods and different sources of data

Yin (2009) advocates the use of triangulation in data collection for a case study. Triangulation was touched upon earlier in this chapter but will be discussed in more detail here.

Triangulation uses multiple sources of data to support a case study. One example Yin (2009) gives is that of researchers collecting oral histories from key personalities featuring in a study. These are later supplemented by more traditional historical sources. The two sources of data complement each other.

As Yin (2009) notes, using evidence from a range of sources allows an investigation to address a broad range of historical and behavioural issues. It permits the development of converging lines of inquiry from a broad spectrum of sources including documents, observations, archives, interviews and surveys.

Patton (2002) in Yin (2009) refers to four types of triangulation including data triangulation and methodological triangulation. Our discussion so far has only looked at data triangulation. Patton (2003) explains how methods triangulation (his version of methodological triangulation) uses qualitative and quantitative data sources as comparators. The two types of data may come up with different findings and these will need to be looked at closely as convergence of findings is the ideal outcome. Patton advocates comparative analysis, that is looking for similarities and patterns between data.

4.2.5 Case study road map

Patton and Applebaum (2003) argue the case for qualitative research methods when conducting case studies. Typically quantitative research has been regarded as a preferable approach having more *'precision, objectivity and rigor'* than qualitative methods. These latter methods are regarded as subjective, open to bias in the researcher, and have a difficulty in generalising from data which is a feature of quantitative research. Patton and Applebaum (2003) propose a schema or 'road map' for conducting *'proper and useful'* case studies.

- **Determine the object of study**

  This is the first and an important step. The researcher must decide the topic of the case study. At this stage it should be broadly defined to allow for amendment and development in new directions. Nonetheless, the aims of the research must be outlined, and hypotheses at least sketched out.

- **Select the case**

  The case must be thoughtfully chosen so that it is relevant to the area of study and will allow the subject to be fully investigated.

- **Build initial theory through a literature review**

  The researcher should visit existing literature to frame his research and set parameters for subsequent research and findings. If the findings in the case fit in with existing literature this should support the case. On the other hand, findings that differ from the existing literature can raise useful questions including whether new theory is needed to explain the findings.

- **Collecting and analysing the data gathering**

  Good data collection requires a systematic approach so the focus should remain on the subject of study at all times. All research should aim to be systematic and planned with this aim in mind.

- **Analysing the data and reaching conclusions**

  It is important to remember the context in which the data is gathered, analysed and conclusions reached so the researcher is not overwhelmed by data which is unanchored to any aims. The context relates back to the literature review and triangulation.

4.3 Cross-sectional studies

Cross-sectional studies usually involve selecting different organisations, or units in different contexts, and investigating how other factors vary across these units (Easterby-Smith *et al*, 2002):

Collis and Hussey (2003) explain cross-sectional studies as being designed to obtain information on variables in different contexts, but at the same time. Normally, different organisations or groups of people are selected and a study is conducted to ascertain how factors differ.

Remenyi *et al* (1998) describe cross-sectional studies in terms of taking a 'snapshot' of a situation in time. This type of research does not attempt to comment on trends or on how situations develop over a period of time.

Cross-sectional studies are conducted when there are constraints of time or resources. The data is collected just once, over a short period of time, before it is analysed and reported (Collis and Hussey, 2003).

Advantages

- They are relatively cost and time efficient (Collis and Hussey, 2003).
- They have the ability to describe, economically, features of large numbers of people or organisations (Easterby-Smith *et al*, 2002).

Disadvantages

- The problem of selection of a sample large enough to be representative of the total population (Collis and Hussey, 2003).
- The studies do not explain *why* correlation exists; only whether it does or does not (Collis and Hussey, 2003; Easterby-Smith *et al*, 2002).
- Difficulty eliminating the external factors (variables) which could possibly have caused the observed correlation (Collis and Hussey, 2003; Easterby-Smith *et al*, 2002).

4.4 Longitudinal studies

As defined by Remenyi *et al* (1998) the term longitudinal is used to describe a study that extends over a substantial period of time and involves studying the changes over the period time.

Collis and Hussey (2003) describe it as a study over time, of variables or group of subjects. The aim is to research the dynamics of the problem by investigating the same situation or people several times, or continuously, over the period in which the problem runs its course. Such studies allow the researcher to examine change processes within a social, economic and political context.

Adams and Schvaneveldt (1991) argue that in the observation of people or events over time, the researcher is able to exercise a measure of control over variables being studied, provided they are not affected by the research process itself.

Longitudinal studies have the following advantages and disadvantages.

Advantages

- There is a capacity to study change and development over time (Saunders *et al*, 2009).
- It takes into account the impact of social process and focuses beyond the individual (Collis and Hussey, 2003).
- Smaller sample sizes may be used and may make access easier (Easterby-Smith *et al*, 2002).

Disadvantages

- The methodology is very time consuming and expensive to conduct (Collis and Hussey, 2003).
- The complexity of data requires very high skills from all researchers involved (Easterby-Smith *et al*, 2002).
- Over time, there is likely to be some drop-out of research subjects (Collis and Hussey, 2003).

The length of time and resources involved in conducting a longitudinal study may mean that it is not an appropriate strategy for a Masters research project except where the investigation is based on secondary data.

4.5 Ethnography

The purpose of ethnography is to interpret the social world the research subjects inhabit in the way in which they interpret it (Saunders *et al*, 2009).

Collis and Hussey (2003) define ethnography as an approach in which the researcher uses socially acquired and shared knowledge to understand the observed patterns of human activity.

The following is a summary of the characteristics of ethnography outlined in Denscombe (2003).

- The researcher spends considerable time in the field among the people whose lives and cultures are being studied.

- Seemingly mundane and routine aspects should be considered as data.

- Focus is placed on how the members of the group/culture being studied understand things.

- There is a holistic approach which stresses processes, relationships, connections and interdependency among the component parts.

- The ethnographer's final account of the group or culture being studied is more than just a description; it is a construction which employs particular writing skills (rhetoric) and which inevitably owes something to the ethnographer's own experiences.

The disadvantages of ethnography as identified by Denscombe (2003) are summarised as follows.

- There is a potential to produce an array of 'pictures' which co-exist, but which tend to remain as separate, isolated stories.

- Descriptive accounts may develop at the expense of analysis contributing to a theoretical position.

- The researcher can get too close to the research subjects thereby obscuring a vision of the obvious.

4.6 Grounded research

According to Collis and Hussey (2003) the purpose of grounded theory is to build theory that is faithful to and which illuminates the area under investigation. The intention is to arrive at prescriptions and policy recommendations with the theory which are likely to be understood and used by those in the situations being studied, and which can be commented upon and, if required, corrected. (Turner, 1981).

Glaser and Strauss (1967) define grounded theory as an inductive, theory-discovery methodology that allows the researcher to develop a theoretical account of the general features of a topic while simultaneously grounding the account in empirical observations or evidence.

Collis and Hussey (2003) explain that

'the theoretical framework is developed by the researcher alternating between inductive and deductive thought. First, the researcher inductively gains information which is apparent in the data collected. Next, a deductive approach is used which allows the researcher to turn away from the data and think rationally about the missing information and form conclusions based on logic. When conclusions have been drawn, the researcher reverts to an inductive approach and tests these tentative hypotheses with existing or new data.

By returning to the data, the deducted suggestions can be supported, refuted or modified. Then supported or modified suggestions can be used to form hypotheses and investigated more fully. It is this inductive/deductive approach and the constant reference to the data which helps ground the theory.'

The following diagram based on a description by Jankowicz (2004) might help to clarify the above grounded theory research process.

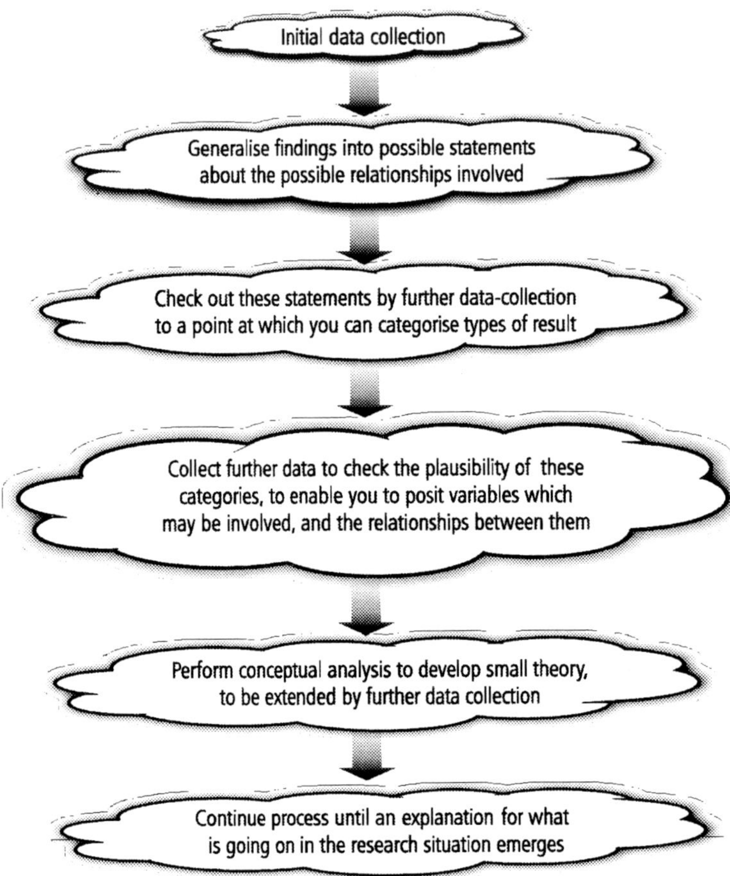

Grounded theory research process

Grounded theory does suffer from some weaknesses.

- The research process generates considerable amounts of data which may be difficult to interpret.

- Views held by the researcher prior to the study may restrict his or her perception of the phenomenon under investigation.

4.7 Experiments

As Marshall (1997) states an experiment is a means of testing the effect of one thing on another, or others.

Saunders *et al* (2009) provide a handy guide to the experimental process.

- Definition of theoretical hypothesis
- Selection of samples of individuals from known populations
- Allocation of samples to different experimental conditions
- Introduction of planned change to one or more of the variables
- Measurement of a small number of the variables
- Control of other variables

From the above it is obvious that an experiment is a form of deductive research.

Denscombe (2003) identified the following, advantages and disadvantages of experiments.

- Repeatability

  - Procedures are carefully recorded and variables controlled.

  - Hence, the experiment may be checked by being repeated by other researchers using identical procedures.

- Precision

  - Precise measurements can be made of the variables that form the basis of the data.

- Convenience

  - Where laboratory experiment is involved, the researcher does not need to spend (too) much time and money.

- Artificial settings

  - The experiment may not replicate conditions existing in the real world.

  - Hence the behaviour of the dependent variable may not reflect what might happen in a real setting.

- Deception and ethics

  - Differences in treatment between the experimental group and the control group may entail an ethical dilemma, eg a hypothesis might propose that improved working conditions improve productivity. It might seem unfair to provide one group with quality office accommodation, new computers and kitchen facilities which are denied to the control group.

  - Also, the ethical propriety of perhaps not fully explaining the purpose of the experiment should be carefully considered. Is deception OK?

- Representativeness of research subjects

  - The experimental group and the control group need to be a 'matched pair' both being equally representative of the population.

  - A practical problem here is 'self selection', ie people put themselves forward and hence determine the composition and character of the groups involved.

- Control of relevant variables

  - This may be difficult to achieve in practice, eg how do you make sure that the bonus scheme is the only thing that affects employee performance? What if the business does well for other reasons and that spurs the experimental group to better performance?

  - Large groups may be needed if the many variations and ambiguities involved in human behaviour are to be controlled. Such large-scale experiments are expensive and time consuming to set up (Bell, 2005).

5 Role of planning

5.1 Acknowledging the task involved

Completing a Research Project is a significant exercise that involves many facets. Moreover, there are also time pressures.

The magnitude and complexity of the task, coupled with the demanding time constraints, are likely to defeat even the most brilliant minds unless proper attention and respect is given to the planning and organisation of the research.

Phillips and Pugh (2005) in their book on completing a PhD stress the importance of planning as a means of overcoming some of the main difficulties inherent in such a large undertaking. Gill and Johnson (2002) support this view in suggesting that for research in general, a lack of systematic planning is likely to be at the heart of many difficulties.

> **GMN viewpoint**
>
> When discussing how to break down the whole task of the research project with students, Ian Crawford offers the following simile.
>
> Question – How do you eat an elephant?
>
> Answer – One bite at a time.
>
> In other words if you consider the entire elephant (the project) in terms of one entire thing to be eaten (completed) then it will become overwhelming. By cutting down into different parts (mouthfuls) and taking one part at a time it is much easier to complete (digest)!

5.2 Importance of process

You will need to apply the best practice used by experienced managers in top businesses in managing your research. In this regard, it is vital to distinguish *process* from *content*. It is helpful conceptually to separate the content of the task from the way the task is accomplished. Based on this analysis, research methods are then primarily concerned with how (process) to tackle tasks (content) (Gill and Johnson, 2002). Hence, a well motivated and self-confident attitude is important but not sufficient by itself in tackling research. Proper planning is vital.

5.2.1 Systematic planning

Bechhofer (1974) suggests that the research process is not a clear cut sequence of procedures following a neat pattern but a messy interaction between the conceptual and empirical world, deduction and induction occurring at the same time. Collis and Hussey (2003) suggest that management research is not a neat and orderly process with one stage leading on to the next. In practice, failure at one stage may mean returning to an earlier stage and many stages are likely to overlap.

5.2.2 Iterative nature

Saunders *et al* (2009) suggest that management research is iterative and may involve revisiting stages of your research (including the research question(s) and objective(s) and working through them again. Support for this approach can be derived from Kolb's experiential learning cycle (Kolb, 1984) which is shown below.

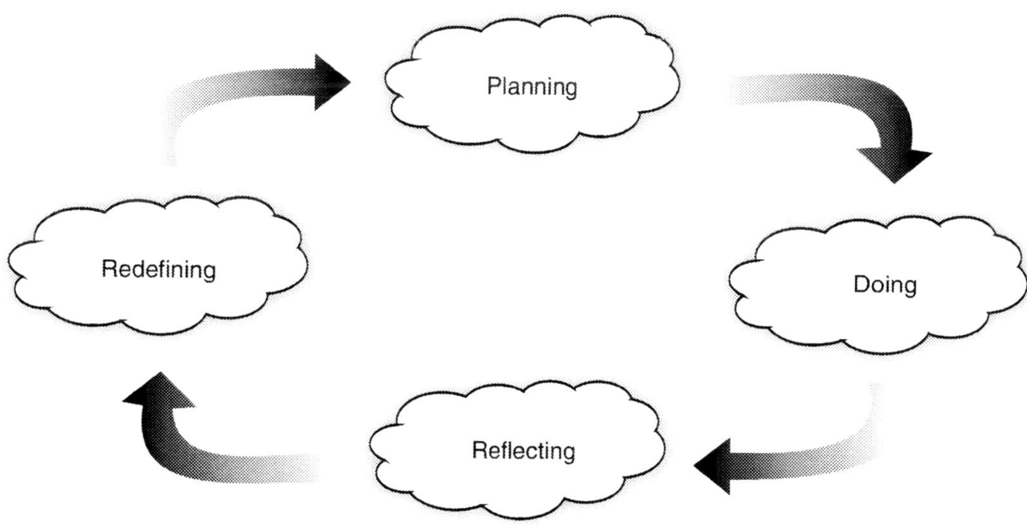

Kolb (1984), adapted

Although the above model allows entry into the cycle at any point, with 'planning' and 'doing' as the recommended starting points, we recommend that you begin with planning. Given the magnitude and complexity of completing a research project it is tempting to resort to a 'dive in and see how it goes' approach. However, the nature of your assignment and the magnitude and complexity of real-life problems you are likely to be investigating means you should begin with careful planning.

5.2.3 Direction and flexibility

The key considerations to bear in mind are likely to be the importance of both 'direction' and 'flexibility' in your work. This is most likely to be achieved by an approach that is based on the planning and iterative approaches outlined above.

- Utilisation of a conscious planned approach as espoused by Phillips and Pugh (2005) and Gill and Johnson (2002)

- Application of an iterative or learning approach in keeping with Bechhofer (1974), Saunders *et al* (2009) and Kolb (1984)

Gill and Johnson (2002) emphasise the importance of planning in determining the efficiency and effectiveness of research work. They suggest that it is especially useful and motivating to be able to identify specific stages of work and be able to agree dates with supervisors. As with all plans, these may need revision from time to time but the presence of an adequate plan enables progress to be assessed at any time. In addition, problems can be foreseen and contingencies can be taken care of.

6 The research process

Collis and Hussey (2003) suggest that whatever type of research or approach is adopted, there are several fundamental stages in the research process which are common to all 'scientifically based' investigations. Saunders *et al* (2009) state that most research textbooks represent research as a multi-stage process that you must follow in order to undertake and complete your investigation.

Assessment advice

We are outlining the entire process here to put the project into context for you but throughout the rest of the module and during the Research Project module we will refer to the various stages in more detail.

6.1 Recommended process

This module provides a research process which is tailored to the requirements of The Global Marketer Programme.

The recommended research process is depicted in the following figure. A brief description of each step is provided here, but fuller coverage will be provided in later chapters. The following diagram outlines the steps to be taken.

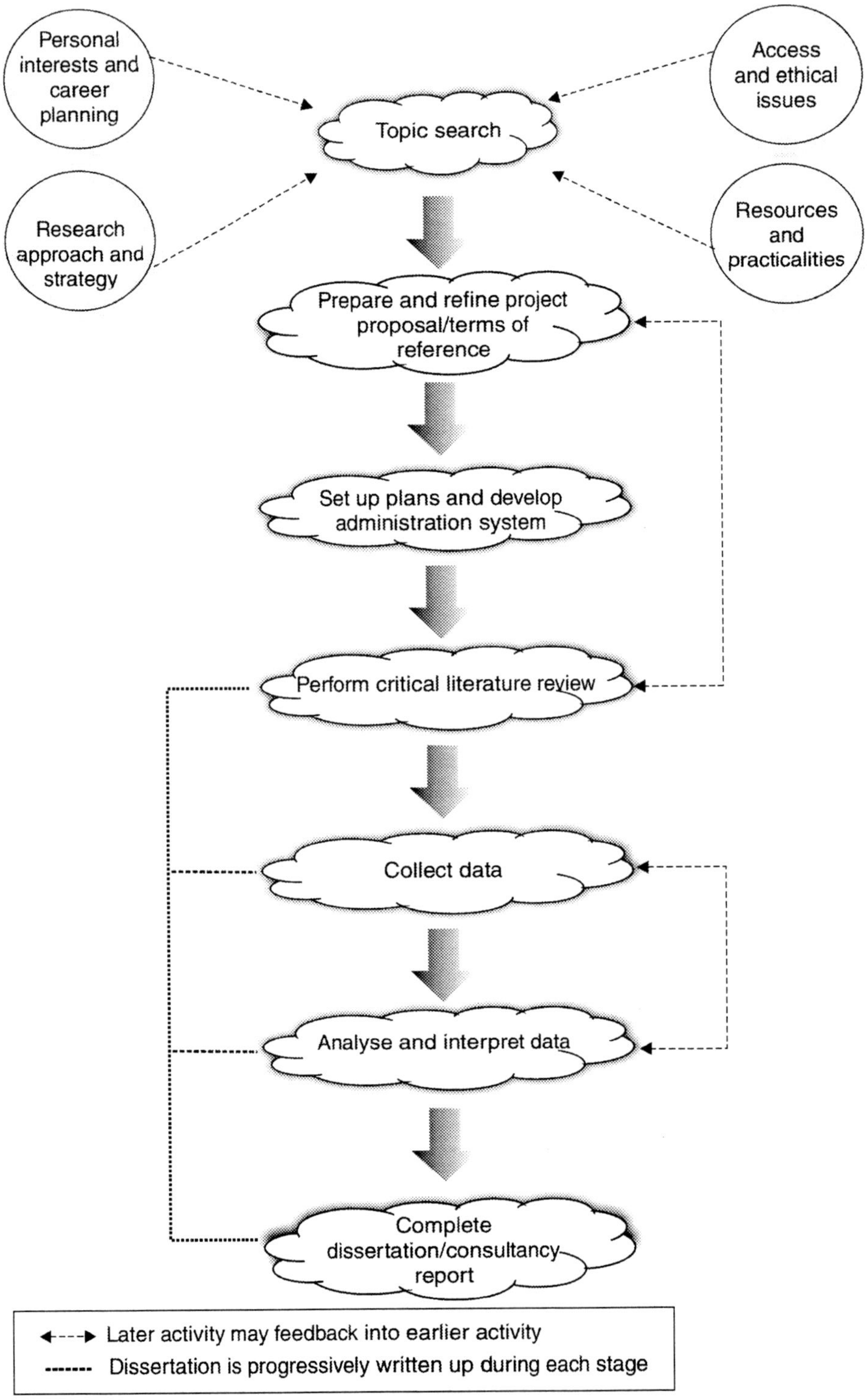

Adapted from Collis and Hussey (2003), Gill and Johnson (2002); Saunders et al (2009)

6.1.1 Topic search

This is a key step in your project as it will determine what you will be doing over the period of your project. You may already have some idea, strong or weak, about what you would like to work on. On the other hand, like many students, you may still have an open mind.

Assessment advice

Chapter 2 will cover in detail how you can select your topic and formulate the research problem.

A key element of this initial stage will in effect be involved in assessing the feasibility of alternative topics. Factors to consider might include personal interest, career relevance, access to information, ethical issues, resources and other practicalities and research approach and method.

6.1.2 Personal interests and career planning

This research is a major piece of work and is likely to involve many academic hurdles and practical frustrations. So it is important that you have sufficient interest in your research topic to sustain your motivation and ensure you complete your investigation.

Careful selection of your research topic could provide you with a springboard in terms of your career planning. Your research could help to upgrade a previous qualification or set of skills, or it may provide a basis for a career move into a new field altogether.

Activity 2

Start to give some detailed consideration of your personal interests from this starting point within the module.

6.1.3 Access and ethical issues

These are very important considerations. Difficulty in obtaining the data you need for your research is likely to seriously hamper the completion of your investigation and prevent you from gaining your Masters degree. So, you must make a thorough and realistic assessment of the chances of getting the data you need.

Assessment advice

Anglia Ruskin University and Global Marketing Network expect you to meet their ethical requirements in completing your project. You must review the research you intend to pursue to ensure that you are working within the university's ethical guidelines at all times.

6.1.4 Resources and practicalities

The perennial considerations of time and money will influence your research topic. You have a deadline to meet and your should ensure that your proposal represents achievable activities. You must be realistic regarding what you can accomplish within the time available.

Research can be costly. Make sure you develop realistic estimates of costs, such as travelling, telephone calls, internet usage, books, stationery and photocopying.

6.1.5 Research approach and strategy

Here you will need to consider what is more appropriate and efficient in relation to the topic you would like to research as well as your own personal preferences and skills.

6.1.6 Proposal preparation

Once you have selected a research topic you would like to pursue, then comes the task of writing your proposal and getting it approved. For this module, the approval process is quite simple. Passing the module means that your proposal has been approved.

You should remember that your proposal is a dynamic statement and might need to be adapted slightly because of something coming out of findings within the research. We will be discussing the research proposal in Chapter 3. We also consider the nature of research design later in this Study Text. Earlier in this chapter we looked at whether you are using a deductive or inductive approach to your research. Depending on which research philosophy you choose will determine whether your research design is fixed or more flexible (Robson, 2009).

6.1.7 Planning and administrative set up

Having had your proposal approved you will be standing on the threshold of doing your research work. As we suggested earlier in the chapter, you should avoid jumping in feet first. Instead, you should develop a clear project plan, which of course you will flex as you progress.

To improve your chances of completing your research successfully and efficiently, there are several areas that are likely to need advance planning.

- Determine logistics for completion
- Negotiate necessary access
- Set up administrative systems
- Establish milestones and timetable
- Prepare personal financial resourcing plans
- Set up physical work space arrangements
- Discuss and negotiate key personal relationships and support

The technical aspects of your research will involve big challenges; you should therefore avoid making your life more complicated because of any personal lack of organisation.

6.1.8 Critical literature review

Saunders *et al* (2009) explain that a *critical review* will form the foundation on which your research is built. Gill and Johnson (2002) suggest that whatever its scale, any research project will necessitate reading about what has been written on the subject and collating it in a *critical review* that demonstrates some awareness of the current state of knowledge on the subject, its limitations and how the proposed research aims to add to what is known.

Hart (2009) provides a warning that a review of a body of literature is often seen as something obvious and a task easily done but this is not actually reality:

'In practice, although research students do produce what are called reviews of the literature, the quality of these varies considerably. Many reviews, in fact, are only thinly disguised annotated bibliographies. Quality means appropriate breadth and depth, rigour and consistency, clarity and brevity, and effective analysis and synthesis; in other words, the use of ideas in the literature to justify the particular approach to the topic, the selection of methods, and the demonstration that this contributes to something new'. p.1

Gill and Johnson (2002) also caution that the literature review takes place early on in the research process and therefore it is important to keep abreast of the literature on the topic throughout the duration of the research. Even with the applied research associated with consultancy, there is a requirement to understand the academic underpinning of your proposed investigation.

Assessment advice

We cover the literature review in Chapter 2.

6.1.9 Data collection

Robson (2009) states that collecting data is about using selected methods of investigation. Doing it properly means using these methods in a systematic and professional fashion.

There are many methods of data collection, because there are many research techniques (Collis and Hussey, 2003). Data can be collected in a variety of ways, in different settings and from different sources (Sekaran, 2002). In research we refer to qualitative or quantitative data. Qualitative data is that which is contextually rich and detailed. Pure qualitative researchers advocate that you should never ascribe numerical values to this data. Quantitative data on the other hand is concerned with larger volumes of numerical data.

6.1.10 Data analysis

It is essential to consider the *validity* and *reliability* of data you intend to use. It is equally important to consider are the appropriateness and suitability of the analysis techniques you decide to use. The analysis tools you employ will depend upon whether you have collected quantitative or qualitative data. Qualitative data tends to be analysed using an approach called Content Analysis. There are many alternative quantitative data analysis techniques which vary from a simple organisation of the data to complex statistical analysis (Robson, 2009).

6.1.11 Dissertation preparation and consultancy reporting

Most textbooks on research methodology describe writing up research as the culminating but separate phase of your dissertation. However, Collis and Hussey (2003) provide the sensible guidance that you should start writing up your research in draft as soon as you start the early stages of your project, and continue to do so until it is completed.

We recommend that, given the flexibility to cut, copy, paste and delete provided by word processing software, you should consider setting up a skeleton of your structure from the outset. You can then make entries directly into your prepared format as you progress knowing that what you have recorded can be edited later on, as and when necessary. The usual file saving and back-up procedures will of course need to be observed. Periodic hard copies might also be useful if your computer were to crash completely or be stolen. The approach should help you to work in an orderly and efficient manner.

7 Differences and similarities between consultancy and academic research

Margerison (2001) noted that the role of the consultant is to assist in improving individual and organisational performance. The consultant must investigate the problems and opportunities that exist, weigh up the options and come up with some advice and information.

Consultancy and pure research are similar as they require a problem-solving approach to investigate organisational activities and then explain them.

Consultancy goes further than pure research as it provides solutions and recommendations. Often consultants are also involved in implementing their recommendations though this is not always the case and depends on what has been agreed between the consultant and the client. However applied research takes the findings of the research and applies them to solving specific problems so here consultancy and research show the same traits.

7.1 The role of the consultant

Margerison (2001) describes four consultancy roles that are used. These roles are the external consultant advising the client, the external project manager advising the client, the internal specialist advising colleagues and the internal manager advising subordinates or colleagues. The roles fit into two dimensions: internal or external consultant and adviser or executive. (Table adapted from Margerison 2001).

External consultant

	A External consultant advising client	B External project manager advising clients	
Adviser	C Internal specialist staff advising colleagues	D Internal manager advising subordinates or colleagues	Executive

Internal consultant

It is clear from these roles that a consultant can be an insider who has an advisory role as well as a brought in consultant.

7.2 Different types of consultancy

Schein (1988) identifies three basic models of consultation. **The purchase of expertise model** involves a situation where the organisation defines a need and recognises that the organisation doesn't have the expertise or resources to fulfil the need. This is when a consultant is engaged to provide the information or service to meet that need. External consultants are often used in this way to plug gaps in the organisation's own resources. The external consultant can also provide 'legitimacy' by being seen as an expert thereby aiding management in whatever changes they are bringing about (Sturdy 2009).

The doctor-patient model uses the consultant as a diagnostician who is engaged to check the organisation over and then recommend solutions to any problems found.

Schein (1988) describes what he regards as some weaknesses in these two models. The success of the expert model depends on a range of factors including whether the manager has accurately identified a need and the consultant has been able to understand the organisation sufficiently to recommend a solution to that problem. The doctor-patient model relies on the consultant being able to accurately diagnose whatever problem exists (if indeed any).

The process consultation model, which is the model made famous by Schein (1988), defines process consultancy as '*a set of activities on the part of the consultant that help the client to perceive, understand, and act upon the process events that occur in the client's environment in order to improve the situation as defined by the client.*' The process consultant works hand-in-hand with the client to diagnose and solve the problem at hand. In doing this the client gains insight particularly into what is happening around him with other people.

Nikolova and Devinney (2007) give another model: the reflective practitioner who conducts the consultancy as a reflective conversation between the client and consultant. By this he means a manager questions and reviews what he has done making the necessary changes to his practice for the next time. This model is time consuming and it is only usually applied when the client's problem is significant.

7.3 Process consultation (Schein)

The process consultation model is one where the client and consultant work together to solve problems. In doing this, the client acquires skills from the consultant and is thereby more able to continue on his own to improve the organisation. The process model involves stages.

Schein (1988) describes these stages as follows:

- Initial contact with the client organisation
- Defining the relationship, psychological contract
- Selecting a setting and method of work
- Diagnostic interventions and data gathering
- Confrontive interventions
- Reducing involvement and termination

7.3.1 Initial contact with the client organisation

Schein (1988) defines this in terms of the client organisation contacting the consultant. The client perceives an organisational problem or lack that can't be solved by current organisational resources.

Usually the exact problem is unclear and so Schein recommends an exploratory meeting with the client to discuss what the real problem is. This meeting can also be used to assess whether the consultant is needed or indeed wishes to take on the assignment.

The meeting can also gauge how willing the client is to engage in joint diagnosis of the problem. Schein suggests an 'open and confronting' approach by the consultant which assesses how open the client is to the consultant coming in and the role of the consultant. Indeed, as Schein (1988) notes, process consulting depends on honesty whereby the client organisation recognises the need for learning in relationships and interpersonal processes. These processes must be willingly opened to scrutiny.

7.3.2 Defining the relationship, psychological contract

Schein (1988) refers to the contract between the consultant and the client as both formal and informal. The formal contract is the scope and outcomes, fees and practical aspects of a contracting relationship.

The informal contract is a 'psychological contract' embodying the client's expectations of what he will gain from the relationship and must give back.

Activity 3

Think about what might form the basis of your formal and informal contract with the organisation you will be completing your research project with.

7.3.3 Selecting a setting and method of work

Schein (1988) advises that the consultant must resolve the setting and method of work in the exploratory meeting or before they begin the work.

The setting should be chosen based on four criteria.

1   What and when to observe need to be worked out in agreement with the client.

     The consultant must concentrate their efforts on observation and feedback where there is joint problem-solving going on between the consultant and those he is observing. So the process involves joint learning rather than the person being observed remotely which may lead to resentment.

2   The setting chosen should be as near the top of the organisation or client system as possible.

     Working at a higher level also bears more influence. Senior managers learning new behaviours are likely to be more influential than at lower levels in the organisation.

     The consultant seeks to work with people in the organisation with the most influence.

3   The setting chosen should be one where it is easy to observe problem-solving, interpersonal and group processes for example a team meeting.

4   The setting chosen should be one in which real work is going on.

     This point is related to the previous one: the members are seen at their 'most natural' but also where the consultant can learn what sort of work the members are most concerned about.

     Schein (1988, p.133) remarks that *'it is much easier to link observations to real work behaviour, and it is much more likely that real changes will occur in members if they can relate process observations to work events'*.

     The **method of work** should make the consultant visible and available for interaction with members of the organisation.

     Schein (1988) recommends interviewing, discussions and direct observation as means of ensuring the consultant remains in contact with the client. Using surveys and questionnaires puts the consultant at a distance from the client and limits interaction.

     Interviewing is useful for establishing relationships with members as much as gathering data.

7.3.4 Diagnostic interventions and data gathering

Schein (1988) observes that once the consultant is busy going about his business of observing, interviewing and attending meetings, he is engaged in diagnostic intervention and confrontive intervention.

Diagnostic intervention runs in tandem with gathering data on the organisation. The process of being observed (data gathering) influences those being studied. Data gathering and the intervention cannot be separate. Schein (1988) remarks that every action by the consultant constitutes an intervention.

So the consultant must try to gather data with this in mind, being 'valid and useful interventions' (p. 142). The only methods of gathering data that would be valid then are direct observation, interviews of individuals and groups and in certain cases questionnaires or surveys. The last method is really only of use where large numbers of people are involved, and to filter out too many choices. The data gathering usually starts off at the top of the organisation interviewing senior people involved in the intervention.

Then further meetings and other forms of data gathering evolve as agreement is obtained with other members to be met.

Issues come up that mean the consultant must 'follow leads' rather than stick to a rigid project plan. The consultant must focus on organisational relationships, processes and organisational effectiveness whatever form the data collection takes.

Schein (1988) also has advice on what style the consultant should adopt during the intervention. He recommends a direct approach, asking questions that mention improvement, effectiveness or whatever is a key issue.

Consultants should avoid being obscure or asking about values and opinions which are open to bias. Asking the right question can lead into other ideas and avenues of exploration.

7.3.5 Confrontive interventions

Schein (1988) defines these as deliberate actions by the consultant intended to change organisational processes. He goes onto explain these interventions in order of their likely use.

- The most likely intervention is one looking at processes and the group's agendas. The group is encouraged to review their internal processes and how they set their agenda.

- Then feedback of observations or other data. The group must be ready for feedback and be receptive.

- Coaching or counselling individuals or groups. This follows on from feedback and can only be used once the consultant is sure the member(s) have really understood feedback and is/are able to start to solve the problem(s) identified.

- Finally structural suggestions on group membership, communication, allocation of work or assignment of responsibility and authority. This can be far reaching as it affects how work is allocated or changes in how communication occurs. The consultant should not be proposing a best solution but helping the client to work one out for himself.

7.3.6 Reducing involvement and termination

Process consultation looks at changing values, if needed, so that the client can use these skills in the future without needing to rely on the consultant. Before ending the intervention the consultant must be sure that they have transferred their skills to the client in interpersonal processes and task management and their values too.

Values that should be adopted include

- Task versus interpersonal concerns. The manager needs to believe that human relations and management of interpersonal and group events are at least as important as task performance.

- Content versus process focus. The manager must be able to give attention to the process by which work is done. This includes considering the personality and feelings of staff when looking at how work is done.

- Short-run output versus long-range effectiveness. Managers must become tolerant of what appears to be slow and calm analysis, leading to long-term improvements in processes and interpersonal relationships.

- Instant solutions versus perpetual rediagnosis. Managers need to be prepared to review processes on an ongoing basis thereby improving task performance. This may be alien to managers who are used to applying instant solutions and believing they will continue to apply.

Of course one of the essential parts of a successful consultancy is the client accepting the outcomes of the consultancy and being willing and able to implement the solution (Schein, 1988 p. 11). Schein remarks that the client must be able to understand the problem and its remedy before they can or are willing to implement the solution. So he regards the process model as overcoming some of the barriers to successful implementation experienced in other consultancy models.

7.4 Examples of different types of consultancy projects

Consultants have a number of roles as you have seen above. Consultancy projects range across a wide range of activities. Consultants may offer a broad range of services or prefer to specialise in certain types of consultancy.

Global case study

The public sector regularly engages consultants to review processes and introduce new systems. Read the following excerpt from a consultancy project between Leeds City Council and a consultant (UPL) involving a Private Finance Initiative (PFI) scheme.

'As part of the Combined Secondary School PFI scheme UPL was employed as a consultancy service to manage and deliver the utilities works required of Leeds City Council as part of the contract. UPL provided an excellent service ensuring Leeds City Council did not incur any additional costs, which often occur due to the complex nature of managing and completing utilities works. UPL provided a professional liaison service between the utilities companies, construction company and the Council to ensure all parties remained well-informed and satisfied.'

Philip Smith: Project Manager.

The extract goes on to explain that the consultant provided project management for a PFI project for the construction of five schools. This included liaison with utility providers to ensure connections and disconnections were maintained and that the project was completed on time.

Source Leeds City Council website accessed 17 June 2009

---

8 Stages of a consultancy investigation

Margerison (2001) gives twelve steps to use in assessing an advisory project. He notes that these may overlap but they are a good guide for checking off the stages reached in a project.

 Research Methods for Global Marketing Practice

1   **Contact**. This is the initial meeting to discuss the problem or opportunity. It involves making an initial contact by the client or consultant. A meeting takes place and the issues are discussed broadly. An agreement is made to meet again if the client is keen and recognises the consultant can help.

2   **Preparation**. The consultant prepares and sketches out their ideas and sends these to the client. In-depth meetings are held to discuss the problem(s). The consultant begins to know the main people at the client's organisation. Margerison terms these the 'actors and their scripts'.

3   **Contracting**. This is the outline proposal of who, when, where, why, what and how much it costs. This is sent as a proposal letter after the parties have met to discuss the contract.

4   **Contract negotiation**. The client and consultant negotiate the details of the contract. The client discusses the proposal with relevant people before agreeing or amending terms prior to the final agreement of the terms of the contract.

5   **Data collection**. Once the contract has been agreed, the consultant starts to gather data by interview, meetings and other sources.

6   **Data analysis and diagnosis**. The data is assessed and discussed with the client. Meetings are arranged to discuss the data.

7   **Data feedback**. Data is presented in a variety of ways usually by presentation, report or both.

8   **Data discussion**. The client and consultant should arrange a meeting to discuss the issues raised by the data in terms of the objectives and purpose. The consultant is usually present to help explain points and avoid misunderstanding.

9   **Proposals**. These should arise from the discussion of the data. Try to keep data discussion separate from putting forward proposals which should follow on. This step is important and all options should be considered and creative ideas allowed expression.

10   **Executive decisions**. The client must come to decisions based on the data. The consultant must not make the decision as it is the client's decision.

11   **Implementation of the decision**. This may be delegated to the consultant or the consultant may advise on this. Only take on implementation if this has been agreed in the contract.

12   **Review**. Ensure you take stock of how the assignment has gone, both in terms of factual outcomes (objectively) and people's feelings (subjectively)

9   Consultancy skills

Block (2000) refers to three sets of skills essential for a good consultant. The first of these is technical skills which include IT expertise or management development. Then, consultants need interpersonal skills for instance listening, negotiation and group or team management. Finally, consulting skills are essential. Block classifies these under contracting, discovery, feedback and decision skills.

In this section, we look in more detail at negotiation, problem-solving, project management and communication skills.

9.1 Negotiation skills

Negotiation is a means of communicating with the aim of reaching a mutual agreement. Negotiation is used for more complex communication where there is some uncertainty over the outcome. The consultant will need to call on negotiation skills at many points in the consultancy. Even when they are tendering for the engagement they may have to debate over points such as price or the scope of the project.

The process of negotiation follows a number of stages which are illustrated below. Much of the negotiation process is actually preparing before the meeting takes place so that you know exactly what you want to achieve and how you intend to do this.

Negotiation involves two main elements:

(a) **Purposeful persuasion**: whereby each party attempts to persuade the other to accept its case by marshalling arguments, backed by factual information and analysis.

(b) **Constructive compromise**: whereby both parties accept the need to move closer toward each other's position, identifying the parameters of common ground within and between their positions, where there is room for concessions to be made while still meeting the needs of both parties.

Such an approach can be applied to a number of different situations.

- **Conflict resolution**: reducing resentment and preserving relationships, by allowing both parties to obtain at least some of their desired outcomes.

- **Group decision-making** and **problem-solving**: integrating different viewpoints and interests so that the decision or solution is high on quality (from diverse relevant input) *and* acceptability (from joint consultation and commitment), enhancing the likelihood of effective implementation.

9.1.1 Approaches to negotiation

There are two basic approaches to negotiation.

Distributive bargaining	Negotiation is about the distribution of finite resources. One party's gain is another's loss: a 'win-lose' or 'zero sum' equation. If a pay increase of, say, 10% is gained, where the management budget was 5%, the extra has to be funded from elsewhere – shareholders (reduced profits), customers (increased prices), other employee benefits (cuts in training) or whatever.
Integrative bargaining	Negotiation is about joint problem-solving, aiming to find a mutually satisfying (or 'win-win') solution to problems. The aim is not just to get the best outcome for one's own party ('win-lose') or even compromise ('lose-lose') but to fulfil the needs of all parties as far as possible.

It is now generally recognised that integrative bargaining is the most constructive, sustainable and ethical approach to negotiation.

Activity 4

Think about your own negotiation skills and how you might improve these in order to maximise the success of your project.

Research Methods for Global Marketing Practice

9.1.2 The negotiation process

The following is a general overview of the negotiation process. Any negotiation will involve the stages illustrated, although the duration and approach of each will vary according to the particular situation.

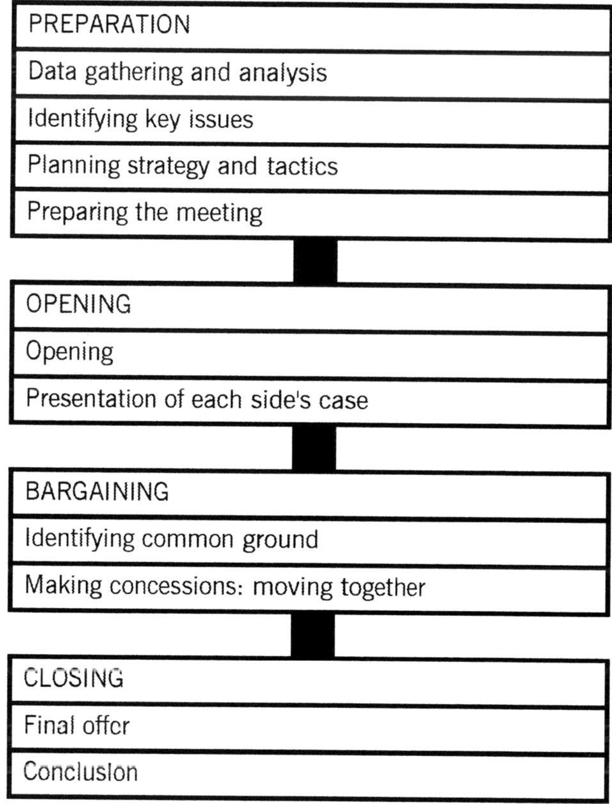

9.1.3 Successful negotiation

John Hunt (1992) lists some characteristics of successful negotiators.

- They avoid direct confrontation.
- They consider a wide range of options.
- They hold back counter proposals rather than responding immediately.
- They use emollient verbal techniques: 'would it be helpful if we...'.
- They summarise on behalf of all involved.
- They advance single arguments insistently and avoid long winded, multiple reason arguments.

9.2 Problem solving skills

In some respects consultants are like fire fighters travelling from problem to problem. However not every fire is the same and often the consultant has less knowledge of the problem faced than the client does. Nonetheless, there are skills that are useful in approaching any problem. We list six steps which can be used to move from the initial stage of diagnosing the problem through to making a decision or coming to a solution.

Margerison (2001) takes this approach further when he refers to 'problem-centred' and 'solution-centred' behaviour. Consultants must spend some time identifying the problem and making a diagnosis before they can prescribe a solution. He advises that it is appropriate to use a problem-centred approach when:

- Problems are open-ended.
- The client must understand how the solution is reached.
- The client is directly involved in managing the situation.
- The solution depends on acceptance of the problem's diagnosis.
- The consultant acts as a catalyst to assist the client in solving the problem.

9.2.1 Solving a problem and coming to a decision

Step 1   Problem recognition

Step 2   Problem definition and structuring

Step 3   Identifying alternative courses of action

Step 4   Making and communicating the decision

Step 5   Implementing the decision

Step 6   Monitoring the effects of the decision

9.2.2 Problem recognition

Decisions are not made without information. The decision-maker needs to be informed of a problem in the first place. This is sometimes referred to as the decision trigger.

9.2.3 Problem definition and structuring

Normally further information is then required. This further information is analysed so that the problem can be defined precisely.

Consider, for example, a company with falling sales. The fall in sales would be the trigger. Further information would be needed to identify where the deficiencies were occurring. The company might discover that sales of product X in area Y are falling, and the problem can be defined as:

'Decline of sales of product X in area Y due to new competitor: how can the decline be reversed?'

One of the purposes of defining the problem is to identify the relationships between the various factors in it, especially if the problem is complex.

9.2.4 Identifying alternative courses of action

Where alternative courses of action are identified, information is needed about the likely effect of each, so they can be assessed.

As a simple example, if our company wishes to review the price of product X in area Y, information will be needed as to the effect of particular price levels on demand for the product. Such information can include external information such as market research (demand at a particular price) and the cost of the product, which can be provided internally.

9.2.5 Making and communicating the decision

The decision is made after review of the information relating to alternatives. However, the decision is useless if it is not communicated. So, in our example, if the marketing director decides to lower the price of product X and institute an intensive advertising campaign, nothing will happen unless the advertising department is informed, and also the manufacturing department, who will have to prepare new packaging showing the lower price.

9.2.6 Implementation of the decision

The decision is then implemented. For large-scale decisions (for example to relocate a factory 100 miles away from its current site), implementation may need substantial planning, detailed information and very clear communication.

Once a decision has been implemented, information is needed so that its effects can be reviewed. For example, if a manufacturing organisation has installed new equipment in anticipation of savings in costs, then information will need to be obtained as to whether these are achieved in practice.

9.3 Project management skills

Some of the skills required by a consultant managing a project are described in the following table. They apply to single-person projects and larger projects involving several consultants and clients. We have included some information on leadership styles where consultants manage teams below.

Type of skill	How the consultant should display the type of skill
Leadership and team building	A participative style of leadership is appropriate for much of most projects, but a more autocratic, decisive style may be required on occasion. Be **positive** (but realistic) about all aspects of the project. Understand where the project fits into the **big picture.** **Delegate** tasks appropriately – and not take on too much personally. Build team spirit through **co-operation** and recognition of achievement. Do not be restrained by organisational structures – a high tolerance for ambiguity (lack of clear-cut authority) will help the project manager.
Organisational	Ensure all project **documentation** is clear and distributed to all who require it. Use project **management tools** to analyse and monitor project progress.
Communication and negotiation	**Listen** to project team members. Use **persuasion** to coerce reluctant team members or stakeholders to support the project. **Negotiate** on funding, timescales, staffing and other resources, quality and disputes. Ensure management is kept **informed** and is never surprised.
Technical	By providing (or at least providing access to) the **technical expertise** and experience needed to manage the project.
Personal qualities	Be **flexible**. Circumstances may develop that require a change in plan. Show **persistence**. Even successful projects will encounter difficulties that require repeated efforts to overcome. Be **creative**. If one method of completing a task proves impractical a new approach may be required. **Patience** is required even in the face of tight deadlines. The 'quick-fix' may eventually cost more time than a more thorough but initially more time-consuming solution.
Problem solving	Only the very simplest projects will be without problems. The project manager must bring a sensible approach to their solution and **delegate** as much responsibility as possible to team members so that they become used to **solving their own problems.** By the nature of a project there is always uncertainty and risk. The project manager needs to be able to react to these situations fast, and adopt an efficient problem solving attitude so as not to hold up the project at key moments.

Type of skill	How the consultant should display the type of skill
Change control and management	Major projects may be accompanied by the kind of far-reaching **change** that has wide-ranging effects on the organisation and its people. Here, however, we are concerned with **changes to the project itself**. Changes can arise from a variety of sources (not least the intended end-users) and have the potential to disrupt the progress of the project. They must be properly authorised, planned and resourced and records kept of their source, impact and authorisation if the project is not to become unmanageable.

9.3.1 Leadership styles and project management

As in other forms of management, there is no 'best' leadership style, as individuals suit and react to different styles in different ways.

The leadership style adopted will affect the way decisions relating to the project are made. Although an autocratic style may prove successful in some situations (eg 'simple' or 'repetitive' projects), a more consultative style has the advantage of making team members feel more a part of the project. This should result in greater **commitment**.

Not all decisions will be made in the same way. For example, decisions that do not have direct consequences for other project personnel may be made with no (or limited) consultation. A **balance** needs to be found between ensuring decisions can be made efficiently, and ensuring adequate consultation.

The type of people that comprise the project team will influence the style adopted. For example, professionals generally dislike being closely supervised and dictated to. Some people however, prefer to follow clear, specific instructions and not have to think for themselves.

Project management techniques encourage **management by exception** by identifying, from the outset, those activities which might threaten successful completion of a project.

9.4 Communication skills

Consultants clearly need good communication skills as they are often in the position of being an outsider bringing in change. Their presence and motives may be misunderstood. Where they work with the client using process consulting the relationship needs clear and close communication in working together to solve problems.

9.4.1 Barriers to effective communication

General problems which can occur in the communication process include:

- **Distortion**: a process by which the meaning of a message is lost 'in translation'. Misunderstanding may arise from technical or ambiguous language, misinterpretation of symbols etc.

- **Noise**: interference in the environment of communication which prevents the message getting through clearly eg due to physical noise, technical interference, or interpersonal differences making communication difficult.

- **Misunderstanding** due to lack of clarity or technical jargon.

- **Non-verbal signs** (gesture, facial expression) contradicting the verbal message.

- Failure to give or to seek **feedback.**

- **'Overload'** – a person being given too much information to digest in the time available.

- **Perceptual selection:** people hearing only what they want to hear in a message.

- **Differences** in social, racial or educational background.

- **Poor communication skills** on the part of sender or recipient.

Depending on the problem, a consultant might consider the following measures to improve communication:

- **Encourage, facilitate and reward** communication. Status and functional barriers (particularly to upward and inter-functional communication) can be minimised by improving opportunities for formal and informal networking and feedback.

- **Give training and guidance** in communication skills, including consideration of recipients, listening, giving feedback and so on.

- **Minimise the potential for misunderstanding**. Make people aware of the difficulties arising from differences in culture and perception, and teach them to consider others' viewpoints.

- **Adapt technology, systems and procedures** to facilitate communication: making it more effective (clear mobile phone reception), faster (laptops for emailing instructions), more consistent (regular reporting routines) and more efficient (reporting by exception).

- **Manage conflict and politics** in the organisation, so that no basic unwillingness exists between units.

- **Establish communication channels and mechanisms** in all directions: regular staff or briefing meetings, house journal or intranet, quality circles and so on. Upward communication should particularly be encouraged, using mechanisms such as inter-unit meetings, suggestion schemes, 'open door' access to managers and regular performance management feedback sessions.

1 Define academic research

- Research is a systematic and not haphazard activity.
- In business research, theory and practice usually go hand in hand.

2 Discuss research philosophies

- There are three levels of theory. Grand, middle level and substantive.
- Epistemology is concerned with the study of knowledge and what we accept as knowledge.
- Positivism focuses on observable and measurable data.
- Phenomenology focuses on social construction and people based meaning founded on personal experiences.
- Inductive research uses mainly qualitative research.
- Deductive research tests hypotheses mostly using quantitative measures.

3 Distinguish between different research designs

- A number of designs were discussed including action research, case studies, experiments, grounded theory, ethnography, cross-sectional and longitudinal studies.

4 Explain the importance of planning to research projects

- Proper planning will help you to complete your research efficiently and successfully.

5 Identify the stages in the research process

- The process involves selecting a topic, establishing administrative systems, reviewing the literature, collecting data, analysing and interpreting the data before completing the report.

6 Identify the differences and similarities between academic and consultancy style research

- Business related research is likely to be of a transdisciplinary nature.
- Management consultancy is likely to be more involved in implementing recommendations and managing change than would be the case in academic business research.

7 Consider the stages of consultancy investigation

- Twelve stages were identified which fit into the broader steps of appraising the problem, assessing the problem and applying solutions.

8 Review the skills required by researchers when working in a consultancy capacity

- A number of skills are required including negotiation, problem solving, project management and communication skills.

1   Phenomenological researchers use qualitative methods to investigate the subjective nature of human activity and its meaning.

2   This is for your own consideration.

3   This will depend on your own organisation.

4   There are many texts on negotiation you can refer to for further assistance.

Adams, G.R. and Schvaneveldt, J.D., (1991). *Understanding Research Methods*. New York: Longman.

Bartholomew, J., (1971). 'The Teacher as Researcher,' *Hard Cheese*, 1.

Bechhofer, F., (1974). *Analysing Qualitative Data*, Routledge: London.

Bell, J., (2005). *Doing Your Research Project: A Guide for First Time Researchers in Education, Health and Social Science*. 4th edition. Buckingham: Open University Press.

Black, T.R., (1993). *Evaluating Social Science Research: an Introduction*. London: Sage.

Block, P., (2000). Flawless Consulting. A Guide to Getting Your Expertise Used, San Francisco: Pfeiffer.

Bryman, A. and Bell, E., (2007). *Business Research Methods*. 2nd edition. Oxford: Oxford University Press.

Collis, J. and Hussey, R., (2003). *Business Research: A Practical Guide for Undergraduate and Postgraduate Students*. 2nd edition. Basingstoke: Macmillan Palgrave.

Cope, E. and Gray, J., (1979). 'Teachers as Researchers: Some Experience of an Alternative Paradigm,' *British Educational Research Journal*. Vol 5, Issue 2 pp.237-251.

Cresswell, J., (2003). *Research Design – Qualitative, Quantitative and Approaches, Mixed Methods*. 2nd edition. London: Sage.

Cunningham, J.B., (1995). 'Strategic Considerations in using Action Research for Improving Personnel Practices,' *Public Personnel Management*, Vol 25.

Denscombe, M., (2003). *The Good Research Guide for Small-Scale Social Research Projects*. 2nd edition. Buckingham: Open University Press.

Easterby-Smith, M., Thorpe, R. and Lowe, A., (2002). *Management Research: An Introduction*. 2nd edition. London: Sage.

Elliot, J., (1991). *Action Research for Educational Change*. Milton Keynes: Open University Press.

Ghauri, P. and Gronhaug, K., (2005). *Research Methods in Business Studies: A Practical Guide*. 3rd edition. Harlow: Prentice Hall.

Gill, J. and Johnson, P., (2002). *Research Methods for Managers*. 3rd edition. London: Sage.

Glaser, B. and Strauss, A., (1967). *The Discovery of Grounded Theory*. Chicago IL: Aldine.

Hart, C., (2009). *Doing a Literature Review: Releasing the Social Science Imagination*. London: Sage.

Hunt, J., (1992). *Managing People at Work: A Manager's Guide to Behaviour in Organizations*, New York: McGraw-Hill Publishing Co.

Husserl, E., (1931). *Ideas: General Introduction to Pure Phenomenology*. Translated by W.R. Boyce Gibson. London: George Allen & Unwin Ltd.

Jankowicz, A.D., (2004). *Business Research Projects* (4th ed.) London: Cengage Learning Business Press.

Kolb, D.A., (1984). *Experimental Learning: Experience as the Source of Learning and Development*. New Jersey: Prentice Hall.

Lewin, K., (1946). 'Action Research and Minority Problems,' *Journal of Social Issues*, 2, pp 34-46.

Luck, M., (1999). *Your Student Research Project*. Aldershot: Gower.

Margerison, C., (2001). *Managerial Consulting Skills A Practical Guide*, Aldershot: Gower.

Marshall, P., (1997). *Research Methods How to design and conduct a successful project*. Oxford: How to Books. Transatlantic publications.

Nikolova, N. and Devinney, T., (2007). 'The Nature of Client-Consultant Interaction: A Critical Review', *UTS: School of Management Working Paper*: No. 2007/3 pp 1-18.

Patton, E. and Applebaum, S., (2003). 'The Case for Case Studies in Management Research' *Management Research News*, Vol 26, No.5 pp60-71.

Phillips E.M. and Pugh, D.S., (2005). *How to get a PhD: A Handbook for Students and their Supervisors*. 4th edition. Buckingham: Open University Press.

Popper, K., (1959). *The Logic of Scientific Discovery*. London: Hutchinson.

Raven, M. and Parker, F., (1981). 'Research in Education and the In-Service Student,' *British Journal of In-Service Education*, 8(1), Autumn. pp42-44.

Reinhart, C.S. & Rallis S.F., (1994). *The Qualitative-Quantitative Debate: New Directors for Program Evaluation*. San Francisco: Jossey-Bass.

Remenyi, D., Williams, B., Money, A. and Swartz, E., (1998). *Doing Research in Business and Management: An Introduction to Process and Method.* London: Sage.

Robson, C., (2009). *Real World Research: A Resource for Social Scientists and Practitioner Researchers* Oxford: Blackwell.

Saunders, M., Lewis, P. and Thornhill, A., (2009). *Research Methods for Business Students* (5th ed.) Harlow: Pearson Education Limited.

Scapens, R.W., (1990). 'Researching Management Accounting Practice: The Role of Case Study Methods,' *British Accounting Review*, vol 22, issue 3.

Schein, E., (1988) Process Consulting Volume 1. Its Role in Organization Development, Reading Massachusetts: Addison-Wesley Publishing Company Inc.Sekaran, U., (2002). *Research Methods for Business: A Skills Building Approach* (4th ed.) New York: Wiley.

Sekaran, U., (2002). *Research Methods for Business: A Skills Building Approach.* 4th edition. New York: Wiley.

Sharp, J.A., Peters, J. and Howard, K., (2003). *The Management of a Student Research Project.* 3rd edition. Milton Keynes: The Open University.

Sturdy, A. Clark, T. Fincham, R. and Handley, K., (2009). 'Between Innovation and Legitimation – Boundaries and Knowledge Flow in Management Consultancy' *Organization*, vol 16, Issue 5 pp 627 – 653.

Somekh, B., (1995). 'The Contribution of Action Research to Development in Social Endeavours,' *British Educational Research Journal*, Vol. 21, issue 3 pp339-355.

Tranfield, D. and Starkey, K., (1998). The Nature, Social Organisation and Promotion of Management Research, *British Journal of Management*, Vol. 9, pp. 341-353.

Turner, B.A., (1981). Some practical aspects of qualitative data analysis: one way of organising the cognitive processes associated with the generation of grounded theory. *Quality and Quantity*, vol15 (3) pp225 247.

Verma, G. K. and Beard, R.M., (1981). *What is Educational Research? Perspectives on Techniques of Research*. Aldershot: Gower. Cited in Bell, J. (2005) *Doing Your Research Project* (4th ed.) Buckingham: Open University Press.

Wheatley, M.J., (1992). *Leadership and the New Science: Learning about Organization from an Orderly Universe*. San Francisco: Berrett Kochler.

Yin, R., (1989). *Case Study Research: Design and Methods* (Rev. ed.) Newbury Park, CA: Sage Publishing.

Yin, R., (2009). *Case Study Research: Design and Methods* (4th ed). Thousand Oaks, CA: Sage Publications

Zikmund, W.G., (2002). *Business Research Methods* (7th ed.) South-Western College Publishers

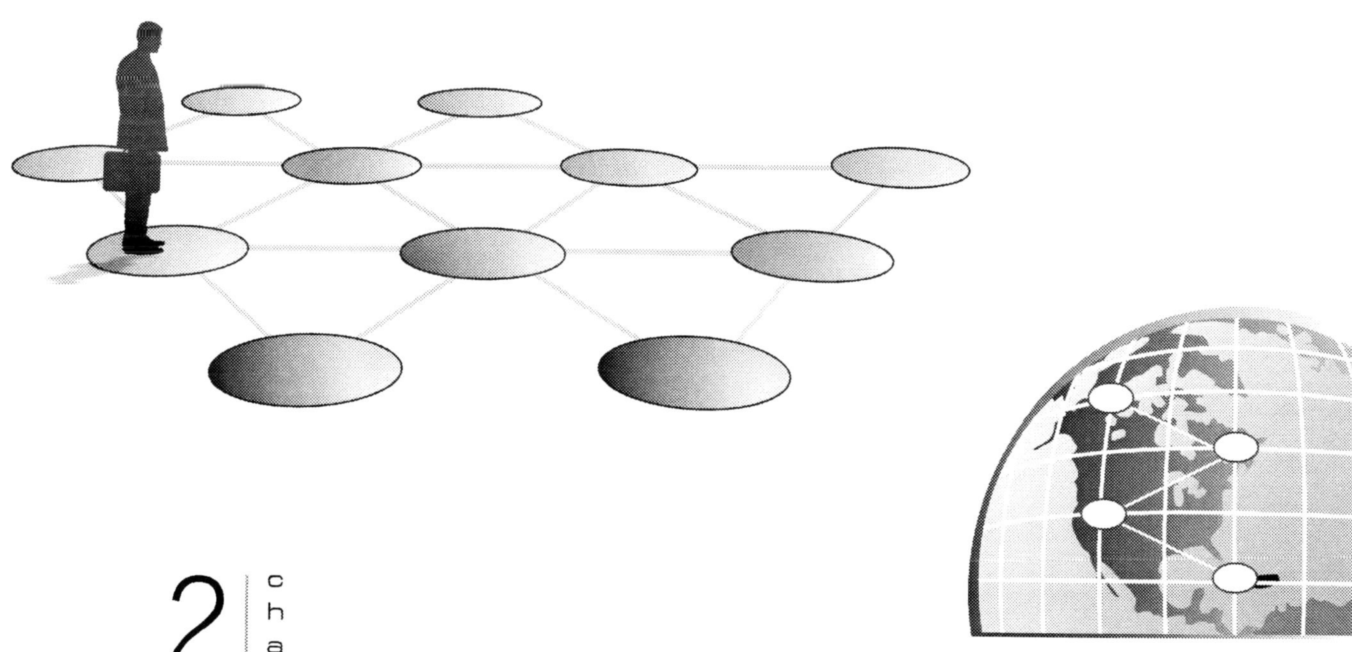

2 chapter

Research problem formulation and reviewing the literature

One of the most important phases in a research project is its definition. This is the point in the project where coherent and clear research aims, objectives and questions to be addressed in the research require careful consideration. As a Masters based research project, you should also ensure that the project is sufficiently detailed to enable you to meet the learning outcomes, includes topics that will retain your interest and at the same time it is worthwhile ensuring that this is also of interest to your organisation.

The chapter begins with practical suggestions for generating ideas for a topic and goes on to look at creative thinking techniques. We then consider the concept of research boundaries before moving to an important stage of your research – the literature review.

Hart (2009) outlines that it is an area where most research students who perform badly do so because of weaknesses in their reviews.

The differences between research questions, objectives and hypotheses are also fundamental issues to appreciate and are covered in the final section of the chapter.

Contents

By the end of this chapter you will be able to:

- Consider how to generate ideas for the Research Project
- Discuss creative thinking techniques
- Explain the importance of scope and boundaries for the investigation
- Discuss the purpose of the literature review
- Identify and evaluate sources of literature
- Plan and conduct your literature search
- Write up your literature review
- Identify the differences between research questions, objectives and hypotheses

1 Generating ideas for a topic

1.1 Sources of information

All sorts of pre-existing material can provide a source of inspiration. Sharp *et al* (2003) suggest the following as useful sources.

- Theses and dissertations
- Articles in academic and professional journals
- Conference proceedings and reports
- Books and book reviews
- Reviews of the field of study
- Communication with experts in the field
- Conversations with potential users of the research findings
- Discussions with colleagues
- The media
- A current organisational issue or challenge

From the list above, you can see that you should not restrict your train of thought merely to books and journals, but should explore some of the numerous other possibilities.

1.2 Past dissertations

Rather than look at individual dissertations for sources of inspiration, you could adopt the approach suggested by Saunders *et al* (2009) and scan a list of past titles. Write down any titles which attract your attention or which could provide the stimulus for further thought, even if you think that the dissertation itself was not very good or the title as stated lacks imagination. We identified a few previous marketing Masters major project titles in Chapter 1.

Assessment advice

The Anglia Ruskin digital library has a number of past dissertations and projects available to view.

1.3 Examination of your own strengths, weaknesses and interests

Luck (1999) believed that being asked to decide upon a subject for your research would either excite you or fill you with dread. In other words, you can view it as a chance to express yourself and broaden your knowledge or you can view it as a threat to expose weakness in your knowledge.

Saunders *et al* (2009) counsel strongly that a research topic should both interest and excite the researcher. Otherwise the researcher will fail to produce his or her best work. They also stress the

importance of selecting a topic in which the researcher is likely to do well and, if possible, have some academic knowledge.

Jankowicz (2004) agrees and suggests that you will have *'less work to do if you choose a topic which you have encountered and worked on before. Analysis, evaluation and judgement will come more easily, and you'll be more aware of the issues that arise if you've covered the basic ground already.'* He advises that the best guide is provided by the marks you have received for the prior work you have done; ie choose something you're good at.

Gill and Johnson (2002) advise that although some researchers may have already defined their topics and, while this may at first sight seem ideal, the work may be impractical or have already been done; either way, the inexperienced researcher may not be aware of the position.

So much for trying to steer you down a 'safety first' route. There is an overriding goal of getting your MSc., so it is indeed tempting to pick a really safe topic. However, you should not forget the importance of self-development and the possibility of building something new that could serve as a stepping stone for your next career move. Jankowicz (2004) recommends that you choose a topic which draws upon your experience but which also helps grow your knowledge or helps you with your next career move.

1.4 Discussions

It is always sensible to talk through any ideas you have for any big project or commitment before you embark upon it. These groups of people can all be invaluable:

- Family
- Friends
- Employer/manager

- Colleagues
- Fellow GMP students
- E-Tutors

As you are to be engaged in the area of business or management research, your employer or manager is likely to be an early and sensible choice.

There may be problems which you have identified or something your manager might like researched. On the other hand, if you are working for a professional service provider such as an advertising agency or marketing consultants, there may be clients who have problems or projects that might benefit from some research. Have a discussion with your line manager to explore possibilities for doing something that might be of mutual benefit to both yourself and the organisation.

Jankowicz (2004), argues that if you are a part-time student following a professional programme, your work experience gives you a great advantage, since your topic is likely to deal with your own organisation. The danger is that you choose a project which you would have been covering already as part of your day-to-day responsibilities. While your project should be relevant, it should also have a measure of academic content.

1.5 Exploratory review of literature

Both academic projects and consultancy projects require a review of literature. We will cover this extensively later in the chapter but it is worth briefly mentioning here because it is useful to conduct as you develop your research proposal.

1.5.1 Sources

As we have seen already, Sharp *et al* (2003) suggest a number of sources within literature. These include:

- Articles in academic and professional journals
- Conference proceedings and reports
- Books and book reviews

The academic content is very important. As stated earlier, the primary purpose of your research is to meet Anglia Ruskin University's requirements for your MSc.

Sharp *et al* (2003) point out that you will need access to a good quality library, which means being 'good' in terms of carrying a wide stock of books and journals, and also in terms of its ability to borrow your required reading from elsewhere. In your case, you will have access to the Oxford Brookes online library facilities.

Note that journal articles and reports (whether of conference proceedings or government reports) tend to be reasonably up to date and are published within a short time of the completion of the work. Books, however, especially in areas of rapid change, such as information technology or the development of management accounting techniques, are sometimes regarded as not being so up to date. They are still invaluable, however, as they can provide a good overview of the current state of affairs in a particular field, and also suggest other sources of information. Book reviews are often of use. Sharp *et al* (2003) recommend that book reviews are useful for students seeking ideas for topics.

1.5.2 What to look for

At this *preliminary search* stage, you are likely to be looking for certain key pieces of information that indicate a *research opportunity:*

- What is known in the field and what is still a matter of conjecture (Sharp *et al*, 2003).

- Unfounded assertions and statements on the absence of research.

- Suitable areas for research and the research methods or approaches that have been traditionally used in this field (Creswell, 2003).

- Contradictions and paradoxes that might suggest that there is an interesting research question to be tackled (Remenyi *et al*, 1998).

You should regard the literature search as an ongoing process. Therefore, as well as being a means of helping to choose a topic, it should also be useful when you are refining your topic.

1.6 Organisationally-generated research ideas

If you are in employment and your employer is sponsoring some or all of your GMP studies, you may have to research a topic specified by the company, or gain your employer's approval for your choice of topic. There are advantages and disadvantages to this, both for the sponsoring organisation and for the student. This is likely to develop into a most practical consultancy approach.

1.6.1 Advantages

Jankowicz (2004) sets out the advantages as far as the sponsoring organisation is concerned:

- The project tackles an issue of relevance to the organisation. This is built into the GMP.

- The project provides an opportunity to examine some issue of corporate or strategic importance, which might otherwise be left low on the list of priorities in the face of competition from more urgent operational issues.

- It can form an inexpensive way of conducting a consultancy exercise.

- If the topic is carefully chosen and refined, it can be of use to the organisation.

- If the student is not permanently based within the company, his or her engagement in research of importance to the company may mean that an employee who would otherwise be engaged in the research can be released to work on something else.

From the student/employee's point of view, the advantages can include:

- Funding and other support provided by the sponsoring organisation.
- Ameliorating any problems of finding a topic or project.
- Providing a platform for a move, presumably upwards, within the organisation.

1.6.2 Pitfalls

As far as the company (or other sponsoring organisation) is concerned, the main deterrents to sponsoring a research project would be:

- The cost
- The lack of any tangible benefit to the company
- The deployment of the individual on non company-related matters

From the individual employee/student's point of view, the pitfalls could include:

- Lack of interest in the company-stipulated project, leading to a loss of motivation

- The research topic being too wide or too narrow or in some other way not appropriate for the requirements of the GMP.

(Saunders *et al*, 2009)

Where there is this conflict between your needs and interests, as the student, and those of the sponsoring organisation, you should try to achieve some form of compromise. If you can motivate yourself by the thought that you are doing something of value to your employer and thereby enhancing your career prospects, this might help to overcome the problem of a lack of great personal interest in the topic itself.

2 Creative thinking techniques

There are many views of what is meant by creativity. It is a word that is freely used in the fields of art, sport and business.

Activity 1

Jot down what you understand by the word creativity. Try to think of an occasion when you were creative. What did you do? Where did your inspiration come from? How did you feel?

According to Boulden (2002), *'creativity is the process of challenging accepted ideas and ways of doing things in order to find new solutions or concepts'*.

2.1 Notebooks

This is probably one of the simplest techniques. All you have to do is keep a notebook available at all times from the point at which you start thinking about your proposal. Jot down any thoughts about topics that occur to you and the source of the idea if that would help. You can then research the idea in greater depth, probably using some of the thinking techniques discussed in this section when you have the opportunity.

There is nothing more frustrating than having a fleeting idea, which you know could be developed further, but then forgetting what it is, or forgetting some subtle twist to an existing idea and being unable to recreate it. Keeping a notebook during the course of your proposal work (and during the course of the work itself) will help to prevent this from happening.

2.2 Relevance trees

A relevance tree enables you to develop ideas from a basic, fairly broad, starting point. You should try to generate three aspects from the original basic idea which then fork out as 'branches'. From each of those a further three branches should fork out, and then a further three from each of those, and so on. Keep doing this until you end up with an idea which you find interesting.

Globalisation is an issue which is currently highly topical, and which many people find extremely interesting. Globalisation as such is so broad that it might be quite difficult to create a relevance tree, but if you narrow the topic down a little it becomes much easier. Restrict yourself to globalisation in the legal and financial sector, and you might come up with something like this.

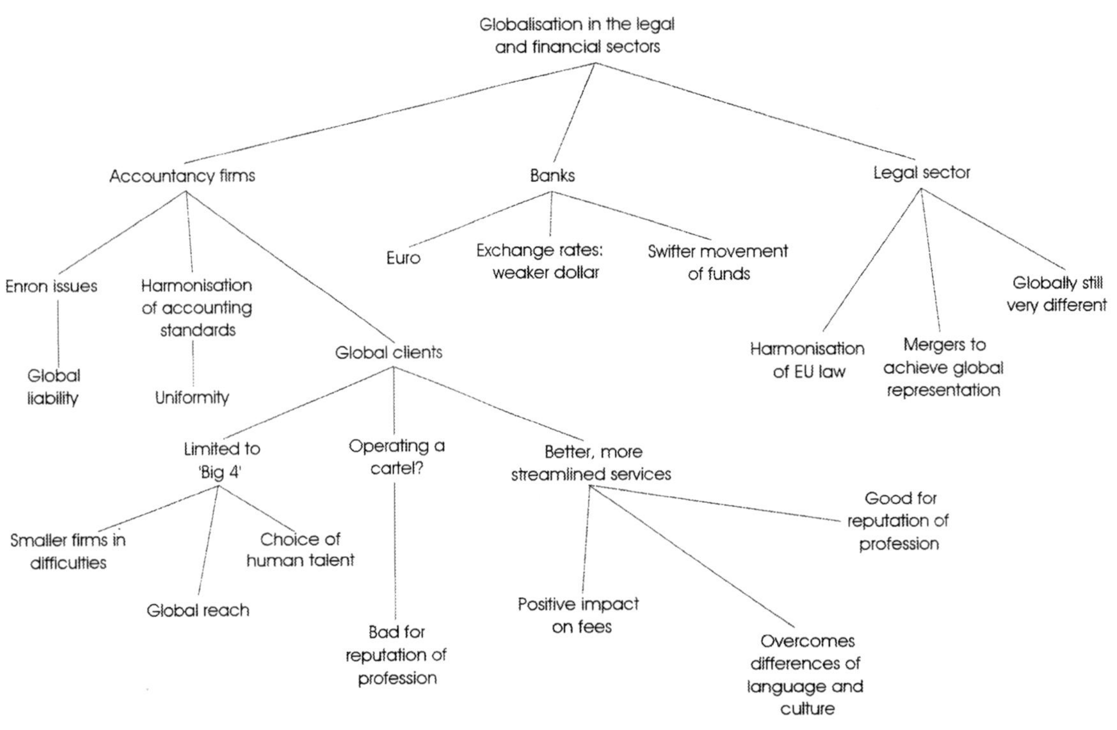

Sharp *et al* (2003) provide an example of a relevance tree based on demand for transport which has only two branches stemming from each idea. It really does not matter whether you have two or more: as long as you allow the tree to branch out in sufficiently diverse directions to enable you to think right across the subject.

Research Methods for Global Marketing Practice

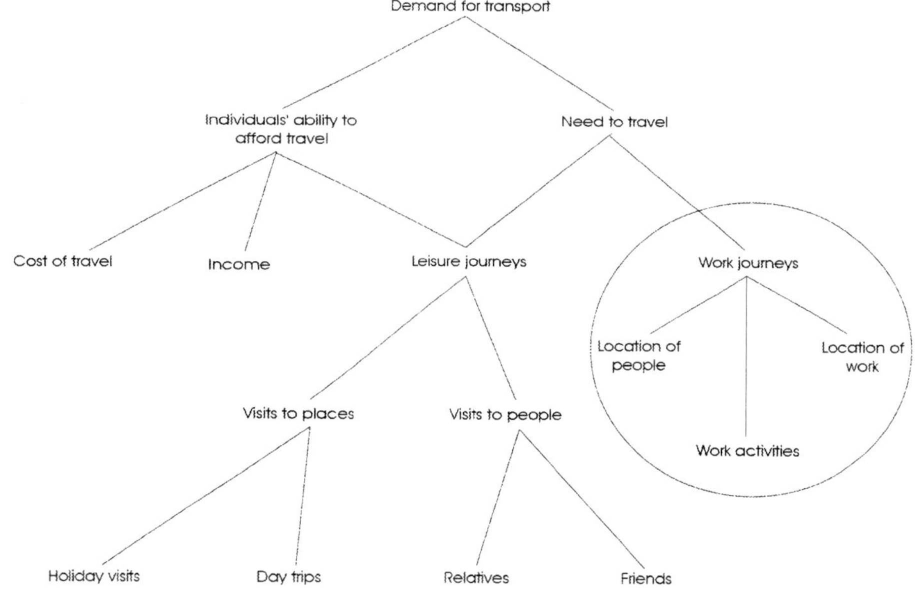

Source: Sharp et al (2003)

2.2.1 Analogy

Sharp *et al* (2003) advocate the use of analogy in what they call topic formation. They suggest that analogy may provide a fruitful line of enquiry on a perceived resemblance to some other area.

Here are some examples of the use of analogy in topic selection.

- Past research found that advanced computer equipment used in the West is inappropriate in countries with a less developed technical infrastructure and to whom intermediate technology is better suited. This indicated a research opportunity relating to the implementation of 'intermediate technology' in British small businesses (Sharp *et al*, 2003).

- Methodology used to study marketing was helped by analogy with the methods used in a study of personnel managers.

2.2.2 Brainstorming

This is a technique which is normally conducted in a group, although Saunders *et al* (2009) suggest that it could be done alone. Performed as a group exercise, it means that you can derive the benefit of the ideas of a group of people. Each person's ideas can then be built upon by the rest of the group. A two hour brainstorming session with six participants can generate far more than twelve hours worth of thinking time, because of the way in which the thought processes of each individual interact with those of the other participants.

Key characteristics of brainstorming are that:

- Group members should be open minded and non-judgemental.
- There should be a free flow of as many ideas as possible.
- Cross-fertilisation of ideas is helpful.

The stages to follow are these:

1   Try to state the problem or the potential topic as clearly and as precisely as possible, although by definition it may at this stage be quite vague.

2   Ask for suggestions from the others.

3   Note down every idea that anyone comes up with, no matter how ridiculous it may initially seem to be. Don't make any judgement or enter into discussion about the ideas at this stage: just concentrate on writing them down.

4   Discuss each of the ideas to seek clarification and consider how it could be implemented as a potential research topic. At this stage you can engage in critical thinking and effectively *'pull the ideas apart'*.

If you are brainstorming alone, the key stages are 1 and 3. You must take the time to state the problems as clearly as you can, and then allow your brain the freedom to wander around the subject. Make a note of anything that springs to mind, but don't start to analyse it at this stage. Allow a little time for your ideas to germinate and then engage in critical thinking, as if you were in a group.

Many practitioners of brainstorming encourage the raising of wild or radical ideas, on the grounds that these are radical but can then be translated into sensible, practical ideas. Of course, in a business you would have to take into consideration any negative publicity if a politically-incorrect suggestion was leaked and published by the media.

Activity 2

You are the manager of a suburban shopping mall which is experiencing severe graffiti problems. It is costly to remove, as well as being unsightly for shoppers. Take a piece of paper and jot down as many ideas as possible. Let your mind roam without any constraints.

2.2.3 Other sources of ideas

Various tried and tested approaches to identifying research topics have been explained above.

Luck (1999) provides a handy and user-friendly checklist that you might find useful, if nothing else, as an *aide-memoire*.

Sources of ideas for projects

*   A TV/radio/newspaper item which caught your attention.

*   Something mentioned in a lecture or tutorial which seemed to be unresolved.

*   A generalisation made by someone outside your field ('Everyone knows that ...'. 'What's the point of ...?', etc.) about which you felt suspicious or which you think could do with testing against the evidence.

*   Something you have always wondered about.

*   A hunch you've had for a while.

*   A quotation that particularly stuck in your mind.

*   A necessity in need of an invention.

*   A flick through a current journal.

*   Previous research projects in the department.

- A chat with a postgraduate about their work (possibly your Study buddy).

- A book or poem you read recently.

- An article that caught your eye in the popular journal section of the library.

- A public lecture you attended (or GMN conference you attended).

- A problem or issue currently facing an organisation.

3 Establishing the scope and boundaries for the investigation

The two most challenging elements in developing your research proposal are:

(a)   The initial generation of ideas
(b)   Turning the idea into a research title

Most of the authorities listed in the recommended reading devote some time to the operation whereby a topic is transformed into a title. You will discover that it is a separate exercise in its own right.

This section aims to help you to make that quantum leap from 'interesting topic which I would like to research' to clearly defined hypothesis or research objective.

Your job now is to turn the general nature of the research idea into something specific that you can focus upon and around which you can design a research programme.

3.1 Refining research topics

The consensus among the authorities on this issue is that you must use your basic research idea to:

- Develop research questions and research objectives (phenomenological paradigm, using either qualitative and/or quantitative data collection techniques)

- Develop hypotheses (positivistic paradigm, using quantitative methods)

Luck (1999) sees it as a process of refining the project's content and thinking about what is feasible, in terms of access to information and ease of research.

Global case study

A researcher wants to investigate some aspect of perceptions of local government and the impact of outsourcing of services to private sector providers. A number of steps would be involved in refining the project.

Step 1

Think through the characteristics of local government, as generally portrayed by the media:

- Delays in service provision
- Poor quality service provision
- Lack of customer responsiveness
- Problems of funding/rising local tax levels
- Increasing bureaucracy
- Staff quality and dedication
- Rigid job demarcations and lack of flexibility
- Lack of opportunities for staff progression
- Uncompetitive salaries and low motivation
- Some success stories

Step 2

On the basis of some aspect that interests the researcher, they pose a research question. For example:

Has the introduction of outsourcing of service through the entering of strategic alliances with private sector companies been effective in terms of improving service efficiency and customer satisfaction?

Step 3

Consider the sources of information:

- Minutes of local government meetings
- Local government newsletters to its constituents
- Published information, including local press records
- Discussions with local government officials and civil servants
- Published performance indicators and benchmarking statistics

Step 4

What about the drawbacks?

- Available information may only be in limited form
- Dealing with very sensitive areas
- Requests for discussions with personnel likely to generate a negative response
- Statistics may be difficult to interpret

Step 5

So how to improve it?

- Retain the elements of the original wording but emphasise the positive, for example:

    'Has the introduction of outsourcing through strategic alliances with the private sector organisations contributed to operational efficiency and customer satisfaction within local government?'

4 Purpose and importance of a critical literature review

4.1 Purpose of search

Brown *et al* (1995) suggest that '*academic research must show that it is based on knowledge of previous relevant work and awareness of the relevant theories, debates and controversies*'. Gill and Johnson (2002) proffer that a literature survey should provide the reader with '*a statement of the state of the art and the major questions and issues in the field under consideration*'.

Business consultancy by its very nature will not attract the same importance to academic literature. However, if you intend to complete a consultancy project as part of your MSc., you will be required to demonstrate how theory and concepts influence the development of your research. This will be part of the reflective essay component of the assessment.

Jankowicz (2004) argues that '*knowledge does not exist in a vacuum, and your work only has value in relation to other people's. Your work and your findings will be significant only to the extent they are the same as, or different from, other people's work and findings*'.

Definition

Sekaran (2002) suggests that a *'literature survey is the documentation of a comprehensive review of the published and unpublished work from secondary sources of data in the areas of specialist interest to the researcher'*.

This is a useful general definition of a **literature review**, though the review should not preclude primary sources of literature.

Collis and Hussey (2003) emphasise that *'the literature review is not merely a piece of writing but an activity which helps guide and inform the research'*.

You are likely to encounter the expressions 'literature search' and 'literature review' in text books. It is useful to recognise the differences between them. Collis and Hussey (2003) distinguish between a · literature search and a literature review as follows.

Literature search: The process of exploring the existing *literature* to ascertain what has been written or otherwise published on a particular subject.

Literature review: A written summary of the finding of a *literature search* which demonstrates that the literature has been located, read and evaluated.

The purpose of a literature search is to help you select, or develop, a framework that will enable you to address your research question. Jankowicz (2002) stated that it involves a *'conceptual analysis of the material you encounter, a critical review of other people's ideas, concepts, research findings, models and theories'*.

4.3 Benefits of literature searching

At the most basic and practical level you are unlikely to want to waste your valuable time researching something that has already been done. So, it is very important to develop a thorough feel for what is contained in the literature. According to Sekaran (2002):

'Sometimes an investigator might spend considerable time and effort to 'discover' something that has already been thoroughly researched. A literature review would prevent such wastage of resources and reinventing the wheel.'

Saunders *et al* (2009) articulate the case for conducting a literature survey most powerfully by stating that *'your critical literature review will form the foundation on which your research is built'*. They explain that the main purpose is to help you develop a good understanding and insight into relevant previous research and the trends that have emerged.

4.4 Learning approach

According to Dell (2005) *'methods used by other researchers may be unsuitable for your purposes, but they may give you ideas about how you might categorise your own data, the ways in which you may be able to draw on the work of other researchers to support or refute your arguments and conclusions'*.

So, you might draw the inference that the process of doing a literature survey helps you to learn and develop an academic writing approach and style which you will need to demonstrate in your investigation.

The various incentives and rationales for carrying out a critical review of literature are well documented in the many books published on research methodology. In light of what has been covered above, have a go at trying to identify, say, ten key reasons for performing a critical review of literature.

Other objectives of a literature review (in addition to those provided in the activity debrief at the end of the chapter) are as follows.

- Demonstrates that the study is empirical.

- Outlines the academic theories within your chosen area.

- Demonstrates up to date knowledge and a competent exploration of the background to the work reviewed relating to your topic.

- Synthesises as well as critically analyses the relevant literature. This should include an assessment of its strengths and weaknesses.

- Justifies your arguments by referencing previous research.

- Through clear referencing, enables those reading your project report to find the original work you cite. This is also a requirement of Anglia Ruskin in their anti-plagiarism rules.

5 Sources of literature

'The different categories of literature resources represent the flow of information from the original source. Often as information flows from primary to secondary to tertiary sources, it becomes less detailed and authoritative but more easily accessible... Some research projects may only access secondary information sources whereas others will necessitate the use of primary sources.' (Saunders et al, 2009).

Recently there have been two UK studies on the persistently annoying problem of the dribbling teapot. As a nation of people who enjoy the traditional 'cuppa' tea, the British had to endure, with their usual stoic fortitude, dribbling from the spout of the teapot after the tea had been poured.

For years there was no progress on solving this technical, yet intractable social problem. Then like the proverbial British bus, two studies came along, almost at the same time! The one presented the vexed problem from a mathematical fluid dynamics approach, building a mathematical model. The other approached the problem from the design of the spout; essentially an applied research approach.

- Professor Jean-Marc Vanden-Broeck of the University of East Anglia discovered that *'the pressure in the fluid underneath the spout is very low. The fluid therefore gets pushed onto the spout by natural atmospheric pressure'*. The professor developed a mathematical formula that is able to predict where and in what shape the dribble would occur.

- In research for a Master's degree in an engineering product design course at the South Bank University, Damini Kumar took Professor Vanden-Broeck's finding a step further by producing a prototype of a non-dribble teapot. The prototype incorporates several features but the principle used to address the dribble problem was to include a rectangular groove on the underside of the

spout So instead of the tea running down the underside of the spout, it would reach the groove and be channelled down into the cup.

Given the social relevance of the findings to many British people, these pieces of research were widely reported in the newspapers and on television.

Now the original papers written by the researchers would represent primary sources of literature, whereas any newspaper reports or video recordings of television coverage would represent secondary data. Often, certain television channels might subject research reports to fairly close scrutiny by a panel of experts and eminent people in the field. Videos or transcripts of these discussions will often provide a critique or scrutiny of the research.

The boundaries between what is a primary source and a secondary source may well be fuzzy because there is no inherently fixed definition of what is which. In practice, you may well find items such as company annual financial reports shown as a primary source in one book and classified as a secondary source in another book. What is important is that your literature review includes what is relevant and useful for your research.

The various publications on research methodology provide detailed lists of their understanding of what comprises primary, secondary and tertiary literature sources.

5.1 Primary sources of literature

Primary sources of literature are the first occurrence of a piece of work and could include many sources. The sources listed are likely to be highly relevant to your dissertation and would be a good starting point.

- Other MSc projects are an obvious source

- Published sources such as reports and some central and local government publications such as white papers and planning documents

- Unpublished manuscript sources such as letters, memos and committee minutes

- Other literature and pamphlets issued by central or local government, eg if your research is the field of local government economic development policies, you are likely to be able to gain access to various documentation on economic development schemes

- In-house company reports and documents such as mission statements, core values statements, procedure manuals, auditors' letters of recommendation, market research projects and management accounts would rank as primary data

(Saunders *et al*, 2009; author's own experience)

Luck (1999) warns that '*getting in to any piece of primary literature can be tough. The style may be unfamiliar and the author may have assumed a lot of background on the part of the reader. It can be especially difficult to overcome the feeling that the paper or article was intended for a small group of*

'experts' of which you are not (yet) one, all of whom understand the jargon and can instantly recognise the significance of the work.'

Although usually masters level research work will involve looking predominantly at tertiary and secondary sources, where you are performing research within your employer or sponsor organisation, you may well include much primary data obtained in-house. The following list of primary sources may provide some inspiration for starting your own search.

- Dissertation and theses
- Project reports
- Unpublished research reports
- Original articles in academic journals
- Research monographs
- Company annual reports
- Management accounts
- Mission statements, core value statements and customer service standards
- Business plans
- Procedures manuals and systems specifications
- In-house market research
- In-house management consulting reports
- Market reports and surveys
- Corporate brochures and booklets
- Auditors' letters of recommendation
- Letters, memos and committee minutes
- Conference reports/papers
- Maps and charts
- Government reports and statistics (some)
- Law reports
- Original material on the internet

5.2 Secondary sources of literature

Secondary sources of literature utilise information already published in primary sources (Saunders *et al*, 2009).

They include sources such as books and journals and subsequent publications of primary literature but these publications are aimed at a wider audience. Secondary sources are also (usually) easier to locate than primary literature as they are better covered by the tertiary literature (which we will be looking at after this section). Many sources are in electronic form and can be accessed via the internet. The following list is not exhaustive but again may give you some inspiration for your own search.

- Books
- Journals
- Newspapers
- Literature reviews in dissertations
- Review articles in research journals
- Review monographs
- Government reports and statistics (some)
- Radio and TV surveys and broadcasts
- Training videos and disks
- Product reviews
- In-house records and systems
- Electronic databases
- Material on internet

Research Methods for Global Marketing Practice

Saunders *et al* (2009) explain that journals, especially refereed academic journals, are the most important literature source for any research. Articles in *refereed academic journals* (such as the *Journal of Marketing*) are evaluated by academic peers prior to publication to assess their quality and suitability. These are usually most useful for research projects as they will contain detailed reports of relevant earlier research.

> ## Assessment advice
>
> Throughout your GMP studies you have been given a journal article to read each week via the Anglia Ruskin digital library.

5.2.2 Practitioner sources of literature

You may also come across journals published by professional bodies. These often contain excellent articles written by distinguished practitioners. However, as Saunders *et al*, (2009) caution, they *'are often of a more practical nature and more related to professional needs than those in academic journals'*.

The articles will often reflect the author's views and comments on issues such as legislation or marketing standards and practices, marketing developments, human resource models and so on, and are not usually based on systematically conducted research. This does not mean that you must not include them in your literature review, but if you do you need to spell out the academic limitations and be able to distinguish between what are either assumptions, assertions, arguments, best practice and researched evidence. They of course may well reveal a wealth of research questions!

Global case study

You might see an article promoting the use of a specific approach, such as the best approach for conducting a marketing performance appraisal. There are many strong arguments in making an intellectual case for what is essentially a best practice. However, there may be no empirical evidence, gained under properly controlled research conditions, that one system of appraisal is better than another.

Perhaps the above exemplifies the paradigm shift that students have to make from the way the professional practitioners such as marketers think, to the way a business researcher thinks.

6 Planning and conducting the literature search

Saunders *et al* (2009) advise that it is important that you plan your literature search carefully *'to ensure that you locate relevant and up-to-date literature'*. They add that all their students found their literature search a time-consuming process which took far longer than expected and suggest that time spent planning will be repaid in time saved when actually searching the literature.

6.1 Planning the search

Luck (1999) warns that *'reading literature for your project will probably demand higher rates of coverage and assimilation than you are used to. It would therefore be wise to make sure you can do it efficiently'*.

Saunders *et al*, (2009) suggest a three step search approach.

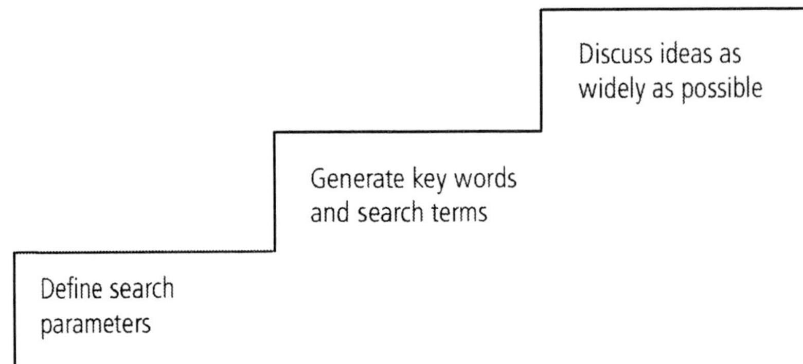

6.1.1 Defining search parameters

Bell (2005) encourages you to identify some search parameters at an early stage, even though these may need to be refined as you go along and actually see the material resulting from your search. Here are some of the key parameters identified in the literature on research methodology:

- **The period to be covered by your research**. Although prospectively your search will need to be ongoing until you complete your research, you will have to decide how far back to look before the material begins to look out of date. This cut off date can of course be flexible and may be redefined as you do your reading (Bell, 2005; Saunders *et al*, 2009; Collis and Hussey, 2003).

- **The geographic boundaries of your material.** If you live, say, in Montevideo you might want to restrict your search to Montevideo city, or the country of Uruguay itself, or perhaps the whole of South America. Alternatively, you might be brave enough to spread your search over material published in the entire Spanish speaking world.

- **The type of literature you would like to include in your search**. Books, journals, theses, government reports, in-house resources, internet resources (Bell, 2005; Saunders *et al*, 2009).

- **Language**. There may be a tendency to limit the search to material written in English. But you might wish to consider material published in other languages (Bell, 2005; Saunders *et al*, 2009).

- **Single or multidisciplinary approach**. Eg the role of training in implementing corporate strategy (Collis and Hussey, 2003).

- **Single discipline, but multi-concept approach**. Eg the role of internal controls in improving shareholder value (Collis and Hussey, 2003).

- **Business sector**. You need to decide how focused you wish your search to be. Will you look at the entire retail sector or will you focus on only one sector such as supermarkets or perhaps clothing retailers (Sanders *et al*, 2009).

Setting research parameters will inevitably need a modicum of common sense. You may well find that the parameters you started with give you either too much literature or too little literature. You may therefore need to experiment with your parameters till you achieve a body of literature which is appropriate to your research.

6.1.2 Using key words

Saunders *et al* (2009) define key words as '*the basic terms that describe your research question(s) and objectives and will be used to describe the tertiary literature*'.

Bell (2005) advises that you should avoid making the focus of your search too narrow, and restricting the number of references you retrieve. To achieve this '*you should organise your topic into subject groups or sets, and analyse the keywords in each to try to find as many relevant search terms (key words) as possible*'. She suggests that this task might be helped by using a thesaurus.

Research Methods for Global Marketing Practice

Brown *et al* (1995) suggest that you should '*play with combinations of key words*'. *They give the example that keyword 'X' might yield 104 references in a search and keyword 'Y' a further 73 references.* Keywords 'X' and 'Y' together may yield only 9 references, which may well be the most important ones for you anyway.

Brown *et al* (1995) also suggest that it may not be easy to find the right keywords straight away and some trial and error may be involved. You need to be flexible and consider other ways in which the topic might be expressed and try alternatives.

6.1.3 Brainstorming, relevance trees and idea sharing

Brainstorming and relevance trees are widely cited in literature as being useful ways of generating key words.

An interactive approach of discussing your work with others can also be helpful. According to Saunders *et al* (2009) 'in discussing your work with others, whether face to face, by electronic mail or by letter, you will be sharing your ideas, getting feedback and obtaining new ideas and approaches'. Likely people include your E-Tutor, your work colleagues, librarians and perhaps even your own friends, if you can persuade them to listen to you!

6.2 Available approaches

There are probably as many approaches to conducting a literature search as there are published books on research methodology.

Saunders *et al* (2009) suggest various approaches.

- Scanning and browsing secondary literature in your library (ARU digital library!).
- Obtaining relevant literature referenced in books and journal articles you have already read.
- Searching using tertiary literature sources.
- Searching using the internet (eg Google Scholar is particularly useful).

6.2.1 Tertiary literature sources

These are also referred to as *search tools* and are designed either to help to locate primary and secondary literature or to introduce a topic. They therefore include indexes and abstracts as well as encyclopaedias and bibliographies (Saunders *et al*, 2009). A comprehensive list of tertiary sources is provided in the list below

- Indexes
- Citation indexes
- Abstracts
- Bibliographies
- Catalogues
- Encyclopaedias
- Dictionaries
- Company search services
- Corporate text search services
- Librarians

Indexes

These index articles from a range of journals and sometimes books, chapters from books, reports, theses, conferences and research.

Citation indexes

These list, by author, the other authors who have cited that author's publication subsequent to their publication. This may provide you with leads to other works that relate to the topic you are examining.

Abstracts

These provide the same information as an index, but also includes a summary of the article. They provide clues to the content of an article you might consider obtaining and reading.

You need to take care not to rely on the abstract itself for your literature review as it might not include sufficient detail for your needs.

Indexes and abstracts are produced in printed and electronic (computerised) formats, the latter often being referred to as databases.

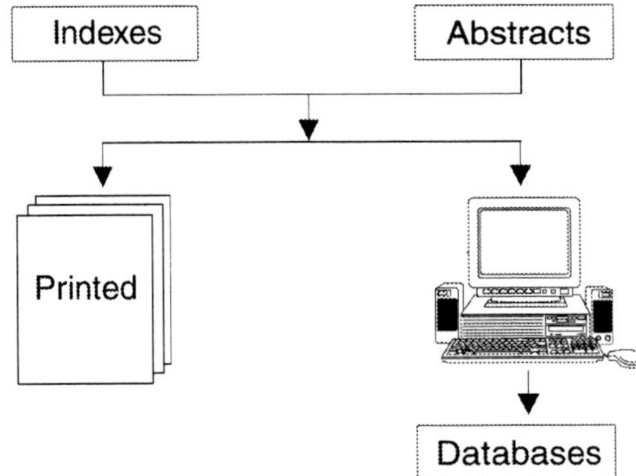

Databases may be in online form or offline, eg in CD-ROM form.

When using databases, your key words must match the database's *controlled index language* of pre-selected terms and phrases or descriptors. Your first stage should be to check your key words with the *index* or *browse* option. Some databases will also have a *thesaurus* which links words in the controlled index language to other terms (Saunders *et al*, 2009).

Some databases now allow *free text searching* of the entire database. This may seem to make life easier, but might lead to an overload of articles and many irrelevant articles.

In searching databases, you can use what are called 'Boolean operators' (ie the words 'and', 'or' and 'not') to combine, limit or widen the variety of items found.

Global case study

We use a neat example provided in Collis and Hussey (2003) to illustrate the use of 'Boolean operators'. Imagine your dissertation topic deals with the marketing of beer and cider in the UK. You decide that your key words are 'marketing', 'beer', 'cider' and 'UK'.

The following table reflects the results of a typical online search using 'Boolean operators'.

Search number	Search words	Number of items
1	Beer	712
2	Cider	45
3	Beer and cider	693
4	Marketing and UK	37,872
5	3 and 4	283

In all searches, you need to be aware of your tactics. A widely defined search will yield a lot of references. A narrowly defined search may yield too few references.

The search for printed sources of literature has various characteristics.

- The coverage of printed indexes tends to be smaller and possibly more specialised than databases.

- It is normally only possible to search by author or by one subject heading, although some cross-references may be included.

- Each issue or annual accumulation must be searched individually, which can be time-consuming.

Books

Luck (1999) suggests that you should look at graduate/postgraduate level textbooks as they often have short bibliographies at the end of each chapter listing key references. However, you should check how recent the publication is. He also suggests you should look along the library shelves for related material as you may find something really useful, albeit serendipitously.

In your case this will be looking at e-book collections on the digital library. You have the opportunity to browse, to see adjacent subject terms, and perhaps discover something relevant by chance.

Remenyi *et al* (1998) provide a formal definition of serendipity as *'the happy or pleasant occurrence of discoveries in the course of investigations designed for another purpose'*.

However, Jankowicz (2004) suggests that 'simple surfing' of the web might not only be fun, but could lead to the discovery of sources that would otherwise not have occurred to you. He however cautions that this can be enormously time consuming!

6.3 Using the internet

The internet may provide some useful material. However, it is not the answer to everyone's information needs. Luck (1999) suggests that *'you may be lucky (!) and find a web site on discussion groups devoted chiefly to your topic. It would be rare for this to tell you much of great research value, but it may help to inform you about where the main developments in your field are taking place. More likely is that you will pick up references to personal research pages put up by students, academics, research groups or libraries in other universities or research centres around the world. Many academics put up their lecture notes on the web and produce interactive computer aided learning (CAL) packages. These sources can give you references to methods, summaries of recent findings and researchers' names to look for. They will also give you a useful feel for the topicality of your subject. You might find some surprising contextual uses of your otherwise narrowly defined research area.'*

Saunders *et al* (2009) caution on the variability of the quality of materials found on the web. They cite an interesting passage by Clausen (1996) who compares the web to:

'... a huge vandalised library where someone has destroyed the catalogue and removed the front matter and indexes from most of the books. In addition thousands of unorganised fragments are added daily by a myriad of cranks, sages and persons with time on their hands to launch their unfiltered messages into cyberspace.'

Global case study

With the widespread use of social media and forums it is likely that you will quite quickly encounter other researchers specialising in your area of investigation.

The most effective way of locating current and up-to-date items is to use a search engine (Saunders *et al*, 2009). '*A search engine trawls the web looking for documents that mention the keywords/index terms you have entered.*' (Bell, 2005). Google Scholar is one of the most powerful for academic purposes.

7 Evaluating literature sources

7.1 Relevance

Bell (2005) argues that '*the aim of your research is to retrieve information of direct relevance to your research and to avoid being side tracked or overloaded with material of a peripheral nature*'. She suggests you should look at the following criteria:

- Respectability of sources.
- Citing of the author's name by others.
- Appearance of author's name in other bibliographical sources.
- Referencing of vital points to facilitate checking.
- References are up-to-date with current developments.

The following checklist for evaluating the relevance of literature can be used.

- How recent is the item?

- Is the item likely to have been superseded?

- Is the context sufficiently different to make it marginal to your research question(s) and objectives?

- Have you seen references to this item (or its author) in other items that were useful?

- Does the item support or contradict your arguments? For either it will probably be worth reading!

- Does the item appear to be biased? Even if it is it may still be relevant to your critical review.

- What are the methodological omissions within the work? Even if there are many it still may be of relevance!

- Is the precision sufficient? Even if it is imprecise it may be the only item you can find and so still be of relevance!

(Saunders et al, 2009; Bell, 2005; Jankowicz, 2004; McNeill, 1990)

Luck (1999) provides a light hearted comment, and perhaps some encouragement, in relation to identifying relevant literature:

'*You are likely to chase several shoals of red herring and many flocks of wild geese before the first part of the hunt is over, but don't dismay. Very often, knowing what you are not looking for is as valuable as finding the correct material straight away.*'

7.2 Sufficiency

Whatever your topic and field of research, there is likely to be a lot of literature available for you to review.

According to Luck (1999) questions that researchers frequently ask can be summarised as follows:

- How long does my literature survey need to be?
- How many references do I need?
- How much is expected?
- When should I stop searching?

Saunders *et al* (2009) suggests that it is *'impossible to read everything as you will never start to write your critical review, let alone your project report'*. They hint that you will know that you have looked at enough when the references you come up with relate to items you have already read.

We reiterate the importance of establishing a good working relationship with your E-tutor who can give you guidance on *'what constitutes an acceptable amount of reading, in terms of both quality and quantity'* (Saunders *et al*, 2009).

Luck (1999) offers the following wisdom:

'One thing to be very sure about is that covering the literature is not a competition – your examiners are looking for an appropriate amount of carefully selected and critically assessed reading and will not at all be impressed by a long list of unstructured material. Remember, the most interesting stamp collections are small, focussed and thematically organised rarities of great value; vast assemblies of random acquisitions are just so many squares of sticking paper.'

8 Writing up the literature review

As you work your way through reading a lot of literature, you will no doubt make lots of notes electronically, on paper or on cards, as you go along. These represent your work in progress and you will have the task of transforming this into an appropriate and academically coherent review of literature.

You will need to demonstrate your academic writing skills and be able to write up your critical review of literature in a scholarly way, with precision and clarity. As you work through your literature review, you will no doubt pick up clues on how it is done, how various researchers write up their work.

8.1 The critical process

Collis and Hussey (2003) explain that *'it is not sufficient merely to describe other research studies which have taken place; you need to appraise critically the contributions of others, and identify trends in research activity and define any areas of weakness'*. They suggest that *'this should enable you to argue that your own research is needed'*.

Sekaran (2002) points out that a *'literature survey should bring together all relevant information in a cogent and logical manner instead of presenting all the studies in chronological order with bits and pieces of uncoordinated information'*

Saunders *et al* (2009) reinforce the above theme in suggesting that *'a common mistake with critical literature reviews is that they become uncritical listings of previous research.'* They advise that within your critical review *'you will need to juxtapose different authors' ideas and form your own opinions and conclusions based on these'*. They reveal that *'the key to writing a critical literature review is therefore to link together different ideas you find in the literature to form a coherent and conclusive argument, which set in context and justify your research'*.

The review should not avoid controversial areas found in the literature; these are likely to enhance the quality of your review rather than detract from it. Also, do not avoid data which is contrary to your own position or views as your critical review must demonstrate academic rigour.

Remember that your review should cover not only the research objectives and conclusions found in the literature, but the research approaches, data collection methods, data analysis techniques and so on adopted by the researcher within the literature. You should assess the validity and reliability of the research undertaken.

8.2 Practicalities

In practice you are likely to be reading through a substantial amount of material. It is therefore important that you develop a sound system of making notes, either electronically or otherwise, to ensure that you

will be able to write-up your literature review in an efficient and effective manner. Ensure you catalogue references meticulously in terms of:

- Bibliographical details
- Brief summary of content
- Supplementary information

In terms of the process of conducting the search, Luck (1999) postulates three golden rules for what you have read:

- Do it properly
- Do it once and only once
- Do it for every paper you read

Gill and Johnson (2002) emphasise the importance 'when embarking on a literature search to ensure that everything that is read is noted systematically at the time. After quite a short period the likelihood of remembering is remote and much time may be wasted at later stages of the research, for example locating a precise reference that has not been recorded when read.'

From the reader's perspective Bell (2005) suggests that the literature review should 'provide the reader with a picture, albeit limited in a short project, of the state of knowledge and the major questions in the subject area being investigated'. From the researcher's perspective, Remenyi et al (1998) suggest that 'by the end of the literature review, the researcher should have a vision of what he wishes to achieve in his or her research'.

The following points, drawn from various sources, suggests the supplementary information which is likely to be helpful in preparing your literature review.

- Useful quotations (in full) with page number

- Strengths and weaknesses of the work

- How the source relates to others ('follows from Smith, 2000', 'repeats Jones, 1999', 'disagrees with Frazer et al, 2001')

- Your current opinion of how useful the source is ('Useful', 'Informative', 'Speculative', 'Critical', 'Peripheral', 'Detailed', 'Rubbish' etc.)

- Methodological details

- Extracts from tables and graphs

- Any contradictions or paradoxes that might suggest research opportunities

(Saunders et al, 2009; Luck, 1999; Remenyi et al, 1998)

9 Research questions, objectives and hypotheses

Once you have begun your literature review you will be in a position to start to develop your research objectives, questions and hypotheses.

A hypothesis is a tentative proposition which is subject to verification through subsequent investigation. It may also be seen as the guide to the researcher in that it depicts and describes the method to be followed in studying the problem. In many cases hypotheses are hunches that the researcher has about the existence of relationship between variables.

Hypotheses, therefore, make statements about relations between variables and provide a guide to the researcher as to how the original hunch might be tested (Bell, 2005).

Robson (2009) agrees that there is a sense in which hypotheses form part of all forms of enquiry. This is the hypothesis as tentative guess, or intuitive hunch, as to what is going on in a situation.

Sharp *et al* (2003) suggest that you should avoid too detailed a specification, limiting yourself to a single hypothesis with considerable potential for testing. If your hypothesis is too broad, you may find that you cannot get to grips with it and it lacks sufficient focus.

Jankowicz (2004) says that the hypothesis is usually, but not always, expressed as a question. This is in contrast to Saunders *et al* (2009), who maintain that a good approach is to frame a general research question to generate a set of research objectives. They prefer objectives to questions as they argue that objectives lead to a greater degree of precision.

As suggested in Chapter 1, hypothesis development is more suitable to a positivistic approach to research where measurement is the researcher's priority, and the research scenario lends itself to the use of quantitative techniques. For example, the question may be 'Does giving more shelf space to the supermarket's own label baked beans increase sales?' This could lead to the development of a hypothesis along the lines of 'increased shelf space given to a supermarket's own label brands, increases sales'. This approach is always subject to the *ceteris paribus* (all other things being equal) assumption, which in research terms means being able to control all other variables (eg special offers, advertising etc) while the research variables are being measured.

On the other hand, the researcher may wish to research the impact of introducing 'people greeters' to a supermarket. (A people greeter's job is to greet people as they enter a supermarket so as to express the store's warm welcome to customers.) Here your research question might be 'Has the introduction of people greeters improved customer satisfaction?' and a research objective might be stated in terms of 'To determine the extent to which the introduction of people greeters has enhanced the quality of customers' shopping experiences'.

Under these circumstances, your research is likely to take a phenomenological inductive route.

Global case study

Here are two examples of how research ideas and questions lead to research objectives or hypothesis. Research idea 1 relates to a phenomenological inductive approach whereas research idea 2 relates to a positivistic deductive approach

Research idea 1

After the 'dot com bubble' some businesses and industries have collapsed spectacularly, while others have survived and indeed become profitable.

Leads to general focus research question

........What is it that the successful dot com companies have done which has caused them to be successful?

Leads to research objective

.........To identify the factors in the establishment of dot com companies in the retail industry which have led to their survival.

Research idea 2

The interaction of discounted accommodation rates and occupancy levels in the hotel industry

Leads to general focus research question

.........What impact does the introduction of discounted accommodation rates have on the occupancy levels of a hotel

Leads to research hypothesis

........The discounting of hotel room rates increase the level of occupancy in a hotel. There is more scope for expensive hotels to reduce their room rates and hence these hotels have a more scope for increasing occupancy levels.

9.1 Research symmetry

Gill and Johnson (2002) define research symmetry as a *'way of reducing the risk in any project'* by trying to *'ensure that whatever the findings from the work, the results will be equally of value'*.

Sharp *et al* (2003) explain that the extent to which the outcomes are of similar value is an indicator of the symmetry of the research. This view is supported by Jankowicz (2004) who refers to symmetry as 'balance' and states that it relates to projects where outcomes are equally valuable whether your expectations are confirmed, or negated. An unbalanced project is one in which evidence that agrees with your expectations, or equally, one in which the evidence that disagrees with your expectations, will have no value.

'Symmetry depends in large part on prior beliefs about a topic which are held within a field of study. If, for example, there is strong support for the view that the eating of sweets damages children's teeth or there is little belief that the phases of the moon affect work output, experiments which confirm strongly held opinion will not be rated highly even though the design and conduct of these experiments cannot be criticised. Obviously, if the research findings were to contradict current belief they would be of potential value but this is unlikely in both cases cited' (Sharp *et al*, 2003).

Saunders *et al* (2009) conclude that without symmetry you may spend a considerable amount of time researching your topic only to find an answer of little importance. Whatever the outcome, you need to ensure you have the scope to write an interesting project report (dissertation).

9.2 The importance of theory in developing hypotheses and in writing research questions and objectives

Much of management and business research is of a practical nature and is often based on specific organisations or industry sectors. However, as we discussed in Chapter 1 this does not mean that your research need not have a theoretical basis. Your project must demonstrate a sound theoretical foundation and involve the exploration, testing or formulation of theory using either a positivistic or phenomenological approach.

Saunders *et al* (2009) maintain that theory is something rooted in our everyday lives, and that every purposive decision that we take is based on theory. For example, if you leave your house at 7.30 am, in theory you should be able to catch the 8.00 am train. Therefore theory is also important in research, as research also involves examining the relationship between two variables and explaining and predicting those relationships. As Gill and Johnson (2002) point out, explanation enables prediction, which in turn enables control. The work you do will test, and possibly develop and amend, theory.

Saunders *et al* (2009) summarise this well by stating that although intelligence gathering will play a part in your research, it is unlikely to be enough. You should be seeking to explain phenomena, to analyse relationships, to compare what is going on in different research settings, to predict outcomes and to generalise; then you will be working at the theoretical level.

9.2.1 Objectives

Setting research objectives is one of the most critical stages in the entire research programme. Research which is based on flawed objectives often merely leads to a finding that more research is required in order to address an underlying problem or issue. For example a company may commission research to find out whether customers would buy a new product. The issue that the company should have really investigated is whether customers would replace what they already buy with the new product. In other words, whether they like the new product enough to change their behaviour.

Objectives in a consultancy setting are usually refined through a series of discussions between the researcher and the client.

Frequently objectives are too broad (not providing sufficient direction) or too narrow (precludes consideration of other courses of action).

(a)   Examples of overly broad objectives are: (1) developing a marketing strategy – this project could be endless (2) improving the competitive position of the firm – from which basis? (3) improving the company image – amongst whom, any ideas how?

(b)   Examples of too narrow objectives could be: (1) decrease the price of the brand to match competitor's price cut – is this really the central issue? (2) to specify whether blue or green should be used for new product packaging – are other colours also appropriate? (3) to outline why customers will buy this product.

In order to ensure that your objectives are clear, and neither too broad or narrow it is good practice to use working **research questions** which follow directly from the objective. The research objectives should flow directly from the overall research aim or rationale, research questions then flow from the objectives.

Research questions should not be confused with fieldwork questions which are used within data collection tools such as surveys and interviews. Research questions are not normally phrased in a way that you could use directly with respondents.

Assessment advice

Research questions are questions that the objective will address stated in very specific terms. They should not be confused with fieldwork questions.

The diagram below outlines the relationship between the different levels of defining the research problem. Please note that to be able to show the relationship more clearly, only the research questions and fieldwork questions for objective 1 are shown. In reality, the diagram would look like a pyramid with one aim at the top which then cascades down to several objectives each of which have several research questions. Each research question will then have several fieldwork questions which may or may not be used within the same data collection tool eg survey, focus group etc.

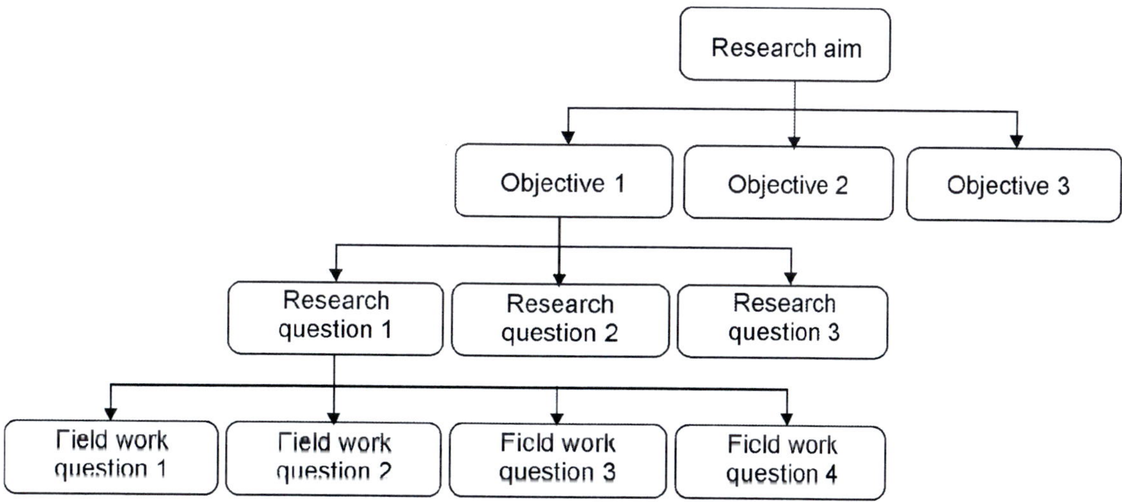

Assessment advice

Given the time constraints within your research project you may only develop one or two objectives.

Let us consider a hypothetical example which may be encountered in a consultancy style project. This example is based on a Spa located in France.

A background for our spa example

Refresh is a small chain of day spas with six spas currently located in Paris, Nice and, Cannes. The company recently acquired an existing chain (Detox Day Spa) and are now undergoing a re-branding exercise to bring the remaining sites under the Refresh brand with equivalent facilities and treatments available in each of the locations.

The spas are 'day only' with no overnight facilities. The main USP (unique selling proposition) is that they are dedicated day spas rather than simply offering treatments within a branded hotel as is the case with many of their competitors.

Currently there are no membership options to join the spas with the focus being on full or half day 'top to toe' packages. Prices differ according to the day of the week with peak periods being Friday, Saturday and Sunday. Monday to Thursday 20% discounts are offered and additional offers are made periodically such as 'bring a friend for free' and specialised talks organised with celebrity health and beauty advisors.

Refresh has been positioned to appeal to women between the ages of 25 – 60 who are interested in their health and beauty regimes but who are looking for relaxation opportunities.

Detox, (the acquired branches) had been a members only spa with a brand positioning which focused on diet and fitness. Over the last four years, Detox lost 30% of their membership base and existing members are aware that Detox will change to the Refresh brand in the long term.

The research will help the Detox marketing team make decisions about:

- Whether membership options should be offered by Refresh
- The new positioning strategy across the chain
- The facilities and treatments that should be offered

Aims, objectives and research questions

Aim – To assist with decision making relating to the rebranding of Detox.

Objective 1 – To identify whether Refresh should offer membership packages.

Research questions following objective 1

(a)   Would the introduction of membership adversely affect the ethos and atmosphere of the day spa?
(b)   Who would be likely to join as members?
(c)   Should the membership packages remain similar to those offered by Detox?
(d)   What would be the effect of removing memberships from existing Detox members?

Objective 2 – To evaluate alternative positioning strategies

(a)   What is the perception of Refresh among existing customers?
(b)   How do Detox members perceive Refresh?
(c)   What positioning would attract new customers into the spa?
(d)   Should Refresh be positioned as a spa exclusively for women?
(e)   Should the diet and fitness image of Detox be blended with the relaxation image of Refresh?
(f)   What would be the impact on Refresh if the existing positioning was modified?

Objective 3 – To suggest appropriate treatment and product mixes

(a)   Should 'top to toe' days and half days remain the core focus?

(b)   What do customers and potential customers like and dislike about the current treatment range?

(c)   What treatments do competitors offer?

(d)   What new treatments are likely to be available in the future?

(e)   What treatments are most popular among spa goers?

(f)   What treatments would both Refresh and Detox customers expect – are their requirements different?

Objective 4 – To review the facilities that should be incorporated

(a)   What aspects of Refresh's facilities are liked and which improvements do customers think could be made?

(b)   What facilities do competitors offer?

(c)   What state of the art facilities will be available in the near future?

(d)   What is the relative importance of different facilities to customers and potential customers?

(e)   Is it important for all sites to have the same facilities?

1 Consider how to generate ideas for the Research Project

- A number of sources of ideas for your project included looking at past projects, completing secondary research and making the most of your contacts to develop your boundaries.

2 Discuss creative thinking techniques

- We discussed the use of notebooks, relevance trees and brainstorming to assist creativity.

3 Explain the importance of scope and boundaries for the investigation

- This is relevant to ensure you can transform an interesting idea into an achievable research project with defined objectives.

4 Discuss the purpose of the literature review

- We discussed a number of reasons stemming from both academic and practical considerations.

- Importantly the literature review guides the research.

5 Identify and evaluate sources of literature

- Primary and secondary sources are available.

- Sources should be evaluated for relevance and sufficiency.

6 Plan and conduct your literature search

- The search should include three steps:

    (i)   Define search parameters.

    (ii)   Generate key words and search terms.

    (iii)   Discuss ideas as widely as possible.

- Both online and offline resources are available.

7 Write up your literature review

- Research should not simply be described but used to identify trends, common arguments and weaknesses in previous work to argue that your own research is needed.

- There are many practical issues to consider including keeping clear and consistent records and ensuring all papers are reviewed meticulously.

8 Identify the differences between research questions, objectives and hypotheses

- Research questions are questions that the objectives will address and are not the same as fieldwork questions.

- A hypothesis is a tentative proposition which is subject to testing.

1 Creativity is the process of finding new concepts or solutions. Your own creativity may come from a personal source. Many researchers find that creative ideas can strike at any time so keep a notebook to hand ready to note ideas down.

2 Here are a few ideas. You may well have many more different ones.

- Install video cameras
- Employ more security guards
- Severely punish the miscreants
- Coat walls with something that cannot be written or drawn on
- Pass law abolishing paint spray tins
- Lock up all children or potential suspects
- Put curtains over offending graffiti
- Install fencing to prevent potential graffitists from getting close enough to walls
- Close down the shopping mall
- Turn the shopping centre into a youth club, police station, art gallery
- Demolish the shopping centre
- Open a graffiti college
- Hold graffiti competitions

You may well have some wild, perhaps not so socially-acceptable ideas. Write them down on the understanding that they can be translated into sensible and workable actions.

3 We have summarised what we believe are the top ten reasons for performing a critical review of literature. How did you get along?

- Provides background to and justification for your research project (dissertation). It is an activity that should also help to 'guide and inform' your research (Collis and Hussey, 2003).

- Helps to refine research question(s) and objectives; ensures these are stated with precision and clarity (Sekaran, 2002).

- Helps avoid repeating research already done (Luck, 1999; Sekaran, 2002).

- Prevents the likelihood of going down 'blind alleys' which others have already encountered. (Luck, 1999).

- Identifies the gaps in current knowledge (Collis and Hussey, 2003; Luck 1999).

- Highlights research opportunities that have been overlooked implicitly in research to date.

- Discovers explicit recommendations for further research.

- Adds value to or provides informed criticism of what has gone before (Luck, 1999).

- Discovers and provides insights into research strategies and methodologies (Cresswell, 2003; Sharp *et al*, 2003).

- Acknowledges and makes proper reference to previous work. This will help to avoid the risks of plagiarism (Luck, 1999; Saunders *et al*, 2009).

Bell, J., (2005). *Doing Your Research Project.* 4th edition. Buckingham: Open University Press.

Boulden, G., (2002). *Thinking Creatively.* London: Dorling Kindersley.

Brown, S., McDowell, L. and Race, P., (1995). *500 Tips for Research Students.* London: Kogan Page.

Collis, J. and Hussey, R., (2003). *Business Research: A Practical Guide for Undergraduate and Postgraduate Students.* 2nd edition. Basingstoke: Macmillan Press.

Cresswell, J., (2003). *Research Design: Quantitative, Qualitative and Approaches Mixed Methods.* 2nd edition. London: Sage.

Gill, J. and Johnson, P., (2002). *Research Methods for Managers.* 3rd edition. London: Sage.

Hart, C., (2009). *Doing a Literature Review: Releasing the Social Science Imagination.* London: Sage.

Jankowicz, A.D., (2004). *Business Research Projects.* 4th edition. London: Thomas Business Press.

Luck, M., (1999). *Your Student Research Project.* Aldershot: Gower.

McNeill, P., (1990). *Research Methods.* 2nd edition. London: Routledge.

Remenyi, D., Williams, B., Money, A., and Swartz, E., (1998). *Doing Research in Business and Management: An Introduction to Process and Method.* London: Sage.

Robson, C., (2009). *Real World Research: A Resource for Social Scientists and Practitioner Researchers* Oxford: Blackwell.

Saunders, M., Lewis, P. and Thornhill, A., (2009). *Research Methods for Business Students.* (5th ed.) Harlow: Pearson Education Limited.

Sekaran, U., (2002). *Research Methods for Business: A Skills Building Approach.* 4th edition. New York: Wiley.

Sharp, J. A., Peters J. and Howard, K., (2003). *The Management of a Student Research Project.* 3rd edition. Buckingham: The Open University Press.

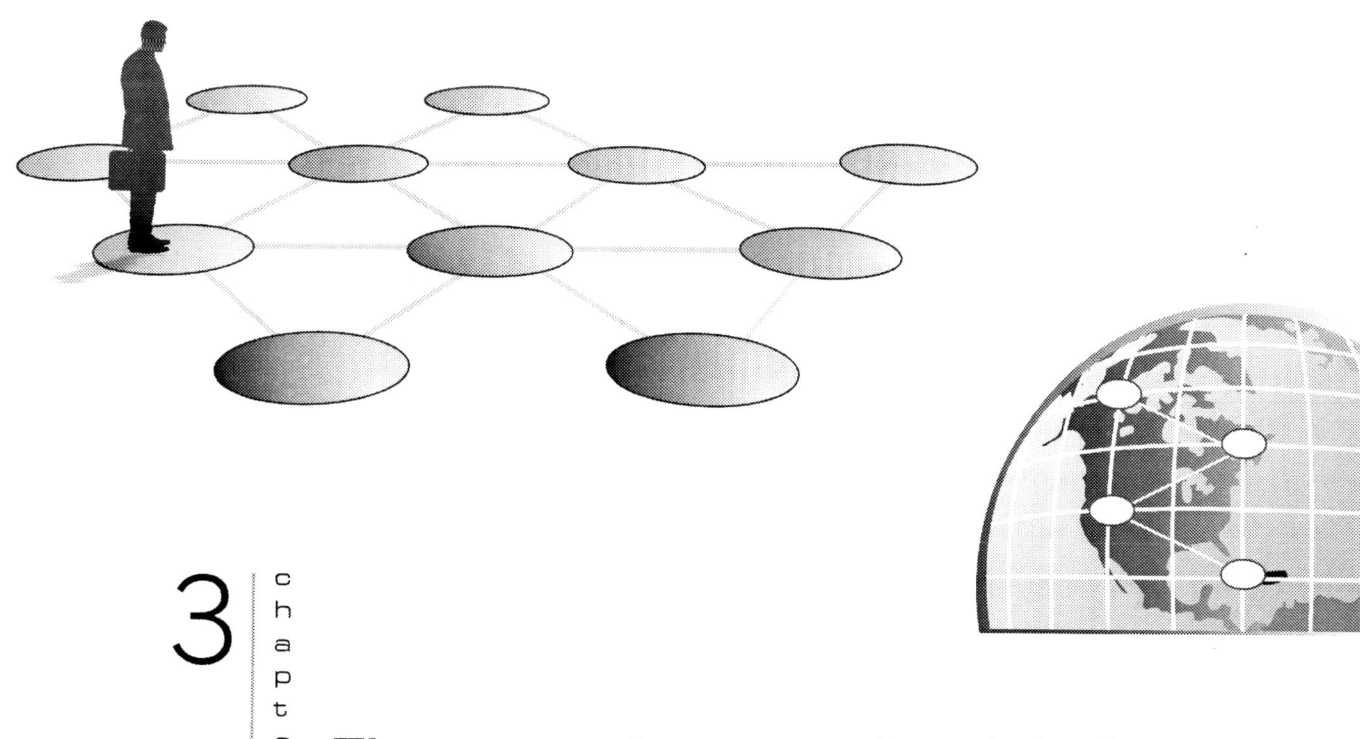

3 | chapter

The research proposal and design

There are various types of research proposal. One is in the form of a research brief created by research agencies or consultants in response to a formal invitation to tender by a client. Differing a little from this, there is also the research proposal created for academic research projects as outlined in Chapter 1. Academic research is empirical in nature and can be based on experience and observation. According to Collis and Hussey (2003) it involves the idea of resolutely looking for the required information.

The overall research design will be 'proposed' within your research proposal and it is worthwhile evaluating the complexity of the overall design and taking a moment to reflect fully on the range of available methods.

Sampling is essential to include in your methodological considerations along with a plans to ensure that your research is both valid and reliable.

In practical terms, management research may often involve a variety of challenges such as gaining access to information and having to deal with the internal politics within an organisation in an ethical manner. We end the chapter looking at these ethical issues.

Contents

By the end of this chapter you will be able to:

- Appreciate the purpose of a research proposal
- Identify components of a research proposal
- Evaluate types of data and their associated properties
- Evaluate the value of mixed methods approaches to research
- Critically review the value of secondary data
- Apply sampling procedures within your research project
- Recognise the importance of validity and reliability in research
- Apply ethical principles in your research

1 The purposes of a research proposal

The first step in the research process is to write a proposal. This is important because it establishes the feasibility of your research in terms of various key aspects.

- You have the resources in terms of time, money and equipment to complete the project.

- You will be able to gain the access necessary to collect the data you need. This might involve having the necessary personal contacts in relevant organisations.

- You have found a suitable research topic that will meet the academic criteria of Anglia Ruskin University.

- The research topic is interesting enough to sustain your motivation and help you complete the project by the deadline.

- Without a properly thought out and well prepared proposal, you are likely to flounder and find it virtually impossible to proceed. The proposal will serve as a handy framework to guide you through the research process.

Assessment advice

You might find it useful to view your proposal as a feasibility study of the research you would like to carry out. It is the document which you will use to persuade your E-Tutor (and examiners) that you have a viable and practicable piece of research. You must put forward a convincing case before you get the go-ahead to proceed.

You will need to pass this proposal in order to proceed onto the Research Project for Global Marketing Practice module.

Bryman and Bell (2007) point out however that a research proposal should be considered a working document and so ideas set out can be refined and developed as research progresses. This is likely to be increasingly true if you adopt an phenomenological perspective.

1.1 Epistemological outlook

Your own epistemological outlook is likely to be a key factor in the type of research that you are likely to enjoy and succeed in. If your outlook is positivistic, you are likely to be more comfortable with deductive, quantitively oriented research, though some mixture of qualitative aspects may well help to give your research a more rounded and deeper feel as well as taking your own self-development into a new dimension. Similarly, a phenomenological approach might derive added value through the application of

some quantitative techniques, if appropriate. In fact the vast majority of research projects employ a mixed methods approach as we shall discuss later.

2 The components of a research proposal

Your proposal should have an academic approach and cover all of the learning outcomes as indicated within the Anglia Ruskin assignment.

Research proposals should take the following format:

Academic research proposal format

Background and academic context

- Including the rationale for the research and identification of critical issues

- The topic should be defined and the organisation/industry context explained

Literature review

- Including significant academic debates and their relevance to the research

Research questions and objectives

- Research objectives and their related questions to be addressed within the research

Research methods

- An explanation of the research strategy and epistemological approach

- Discussion of specific primary data collection tools (including phases of research, types of data required, data collection tools, sampling plan for each data collection method and a brief plan for data analysis)

- Sources of secondary data

- Research ethics

- Access arrangements and contextual hurdles to overcome

Timescale and resources

- A clear outline of timings and key stages in the project

- Resource requirements and availability

Now we have outlined what it is that you need to cover within your research proposal we can begin to look at the planning of the overall research design and methodological choices. The broad areas within the research methods section will be discussed hereafter within the rest of the chapter. Coverage of specific data collection tools and analysis however will be covered in later chapters (Chapter 4 for qualitative data and Chapter 5 for quantitative data).

3 Types of data

In Chapter 1 we considered the differences between qualitative and quantitative variables used within research. Chapter 2 looked very briefly at the differences between primary and secondary data.

These are the main distinctions within the alternative forms of data and the research methods you employ will determine which form of data you are able to collect and later analyse.

Definitions

Qualitative research generates contextually rich data. It is generally unstructured and only a small number of carefully selected individuals are used to produce non-quantifiable insights.

Quantitative research is highly structured research conducted using a large sample of respondents to provide quantifiable insights.

Primary data is collected for the purposes of the research project.

Secondary data already exists and has been collected for another purpose.

The following diagram outlines the broad classification of data types.

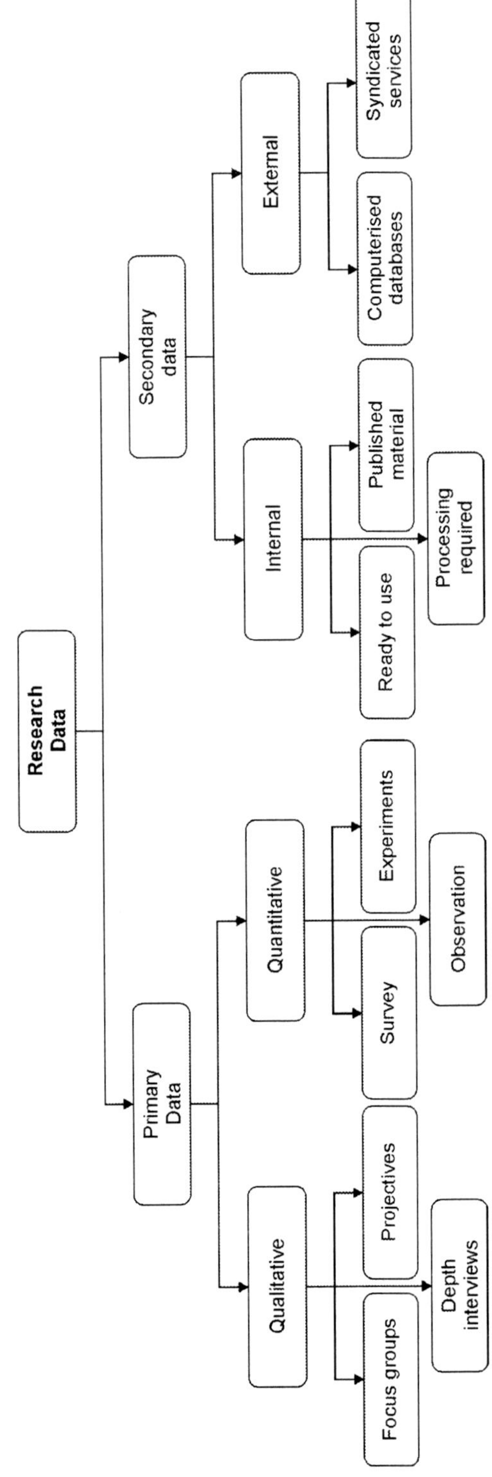

Research Methods for Global Marketing Practice

4 Mixed methods approaches

The mixed methods approach combines secondary data and primary data collected using a variety of methods.

This approach helps to mitigate one of the main weaknesses of using secondary data, because some of it is often not of direct relevance to your research objective. Therefore, you collect appropriate primary data to 'plug the gaps' and assist you to obtain the complete picture. This often happens where you are researching an issue within an organisation.

Bell (2005) advises that efforts should be made to cross-check findings by the use of different methods which then lead you to a consistent conclusion. This is called **triangulation**, which we touched on in Chapter 1.

Multiple methods are of particular value when a controversial area needs to be evaluated more fully than other areas of less difficulty (Cohen and Manion, 1994). They suggest as an example a study into schools looking not just at academic results but also at other areas of potential achievement such as teaching methods, practical skills, cultural interests, social skills, interpersonal relationships and community spirit. Looking at all these factors, they argue, would give the researcher a much more well rounded picture, and the results within each area would hopefully add to the confidence gained in the results as a whole.

Global case study

Imagine you are carrying out research into the effectiveness of different management structures. You would like to confirm a research objective that an organic structure with a team working and information sharing culture is more effective for a publisher of financial and business books than a mechanistic structure, based on classical management principles with strong demarcations of responsibility and individual retention of information, ie information is effectively kept in silos rather than pooled.

As well as researching and analysing pure profitability, you could consider researching other areas. These could include:

- Employee motivation
- Staff development
- Staff turnover
- Teamwork and communication
- Customer focus

- Marketing message
- Customer service standards
- Vision and mission
- Core values
- Corporate culture

You could adopt a range of data collection methods including interviewing key personnel, review of market research carried out and an investigation into the management structures of other similar companies within the industry.

As a general approach, Bell (2005) suggests that you adopt what she describes as a *'problem-oriented approach'*. You would read secondary sources to understand what has already been discovered about the subject, helping to establish the focus of your work, before then proceeding to the relevant primary sources, which would help you to move your research a stage further on, possibly to the point of developing new theory

5 The value of secondary data

The gathering of secondary data can have much in common with the literature search (Sharp *et al*, 2003).

5.1 Advantages of secondary data

Here are the key advantages of secondary data as a source of research information over primary data.

5.1.1 Cost-effectiveness

- Collecting primary data yourself can be prohibitively expensive (for example, the postage on sending out a questionnaire).

- Collecting secondary data is usually more cost-effective (for example, in the context of collecting very wide-ranging or large-scale data, such as that included in official statistics).

5.1.2 Time

- The collection of primary data can be very time-consuming. Even the distribution, chasing up and subsequent analysis of the results of a questionnaire can take many months.

- The use of secondary data from other sources may therefore be a more efficient alternative. This might enable you to use a wider range of data and spend more time analysing and interpreting it.

(Saunders *et al*, 2009).

5.1.3 Quality

- Secondary data, such as, government department statistics will often be of much higher quality than that you could expect to gather yourself (Stewart and Kamins, 1993).

- The collector of the secondary data is likely to have been able to devote far more time and resource (especially financial resource) to the exercise than you.

- Secondary data can provide you with material collected over a long period of time, which you are not in a position to gather yourself.

Global case study

A student researcher would never be in a position to gather as much demographic information as can be found in a document such as the government census.

5.1.4 Discretion

- In using secondary data within an organisational context, it will not be immediately obvious what you are researching, and it will therefore be far less obtrusive.

- Discretion may be helpful if you are researching a sensitive topic such as employee motivation.

Research Methods for Global Marketing Practice

5.1.5 Use as a benchmark

- Secondary data can be useful to compare with your own primary data, and to use as a benchmark. It can be used to put your own primary research into context and to highlight any unexpected discrepancies.

- Secondary data could alert you to potential problems within your own primary research if its results don't stack up with the results of a further-reaching survey.

5.1.6 Unexpected discoveries

- An important research discovery might be made as a result of research into something completely different. An illustration is the inadvertent discovery of penicillin by Fleming, who was actually in the throes of investigating something completely different. It is often the case that new medicines or medical principles are discovered as a by-product of different research.

- Within a management context, the Hawthorne Effect, which highlights the impact of the researcher or the behaviour of participants, was discovered as a by-product of research into factory lighting on worker performance.

5.1.7 A permanent source of information

- Secondary data is permanent (as it has been published) and is available in a form that can be accessed and checked reasonably easily by others (Denscombe, 2003).

- The data and your research findings are more open to public scrutiny (Saunders *et al*, 2009) meaning that they inspire greater confidence in those relying on the data.

5.2 Disadvantages of secondary data

You should be aware of the potential disadvantages of secondary data. Some of them will be of particular relevance when you are evaluating the secondary data you have collected, which will be discussed later on in this chapter.

5.2.1 Access

- Where the original collector of the secondary data has incurred a lot of expense in gathering it, it is less likely that they will provide it to you for free! Large-scale market research reports, for example, may cost hundreds of pounds.

- You may be able to access documents such as these in a library, but according to Saunders *et al* (2009) such material is often barred from the inter library loan service, meaning that you would have to travel a long way to see a library copy.

5.2.2 Relevance

- By its very definition, secondary data has not been initially collected by or for you. Someone else has collected it for his or her specific purpose, under conditions unknown to you.

- As secondary data is unlikely to exactly match your research objective, and you will have to engage in additional, possibly primary, research to complete the picture.

- The secondary data may be now old and the information is not sufficiently current. You might have to engage in primary research to bring the data up to date (Saunders *et al*, 2009).

- The document you are seeking to rely on will have been the subject of a selective approach as to what was included in it, and there may be insufficient supporting data (for example, internal organisation documents or newspaper reports).

5.2.3 Reliability

- It may not be apparent who collected the secondary data and for what purpose. If you do not know whose research it is, it may be unwise to rely on it without carrying out some corroborative research to verify it.

- The authority of the source of the secondary data and the procedures used to produce the original data will need to be evaluated in order to establish its credibility as a source of information for you (Denscombe, 2003).

- If you do not know how the original collector of the data selected a sample, you might not therefore be confident that the sample was representative of the population as a whole.

- The preparer of the secondary data may have been biased or have had some personal or corporate objectives. Trade associations, for example, may not include data that run counter to the interests of their members.

5.3 Types of secondary data

'Secondary data include both quantitative and qualitative data and can be used in both descriptive and explanatory research.' (Saunders et al, 2009). Additionally the data may be described as raw in that it has not been processed to any degree, or it may be described as compiled in that it has undergone some form of selection or summarising.

Saunders *et al* (2009) have drawn on various works on the classification of secondary data, and distilled these into three main sub-groups of secondary data:

- Documentary data
- Survey-based data
- Multiple source data

They are summarised in the following diagram.

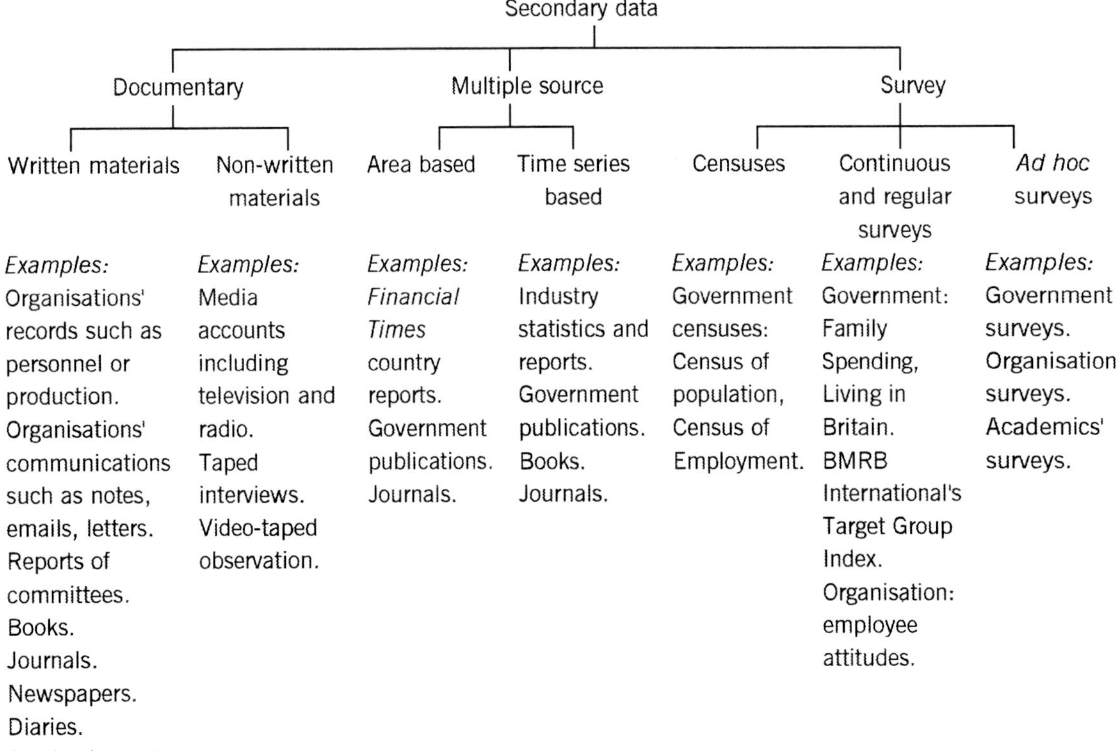

Documentary secondary data

A researcher is investigating the impact of customer service standards in a private sector exam training company on the company's corporate image.

The researcher intends to collect primary data using a questionnaire to survey students and measure performance in respect of the various service standards.

However the researcher is able to have access to various internal company records such as:

- Computerised records of telephone answer response times. One of the service standards is that telephone calls must be answered within three rings.

- Action points from company meetings discussing its performance in achieving its service standards.

- Routine feedback forms collected from students.

- Videos of customer focus group meetings conducted by the company to elicit feedback on its service standards.

The researcher plans to utilise the primary and secondary data to develop a set of findings in relation to its research objectives and research questions for the research.

Multiple source secondary data

A student is researching the impact of foreign investment on the economic development of China.

The researcher intends to collect primary data by interviewing government officials and economists in China.

To supplement the primary data, the researcher might also find various secondary sources to be useful in arriving at the dissertation's findings.

- Country reports by reputable media organisations such as the *Financial Times, The Economist* and CNN.

- Press reports on the impact of foreign investment on economic development in China.

- Available reports and statistics compiled by central and local government bodies in China.

Survey based secondary data

A student works for a leisure company which runs various operations such as ten-pin bowling, roller-skating and associated restaurants. The student is being sponsored by the company to carry out some research on consumer behaviour in the leisure industry with the objective of identifying strategic options which might help to grow the company and enhance shareholder wealth.

In terms of gathering primary data, the researcher intends to administer a survey using postal questionnaires to identify responses to a range of relevant questions. This will be supplemented by focus group work to try to ascertain the underlying drivers of leisure consumers' behaviour. To help provide the primary research with an overall perspective, as well as provide a benchmark for the student's own data, reference will also be made to secondary data, in the form of the Henley Centre's Leisure Tracking Survey.

5.4 Sources of secondary data

Most of the recommended textbooks suggest a wide range of sources of data, some in general terms and some in more specific terms, for example providing lists of addresses of potentially helpful websites.

Sources of secondary data may typically be split into these categories:

- Books and journals
- Technical publications
- Official publications
- Trade association data
- Computer databases

We shall consider each of these in more detail in the next section.

5.4.1 Books and journals

Denscombe (2003) suggests that you should make books and journals your first port of call:

'...they contain the accumulated wisdom on which the research project should build, and also the latest cutting-edge ideas which can shape the direction of the research.'

The university's virtual library may hold much key information as well as pointers to sources of further information.

When evaluating books and their publishers, their relevance and reliability must be evaluated in the same way as for any other form of secondary data.

Denscombe (2003) suggests that you ask yourself these key questions:

1 Have you heard of the publisher before? If you have, it might be reassuring.

2 Is the publisher a university press, such as Oxford University Press? If it is, this could offer some confidence about the academic quality of the work.

3 Is the book in a second or subsequent edition, or has it been reprinted a number of times? If it is, it means that the first edition was sufficiently in demand for the book to warrant new editions or reprints. You will be able to find these details right at the front of the book, usually on the second or third page.

4 If it is a library book, has it been borrowed a lot? If it has, it probably means that the book has been recommended by lecturers or other researchers, and may therefore be good.

The Anglia Ruskin University digital library contains a multitude of journals. Denscombe (2003) offers some similar rules of thumb when evaluating journals as a source of secondary data. The key questions here are:

1 How long has the journal been in existence? If it has been published for some time, you can probably have a reasonable degree of confidence in it.

2 Does the title of the journal include some national element, such as the *British Medical Journal* or the *American Journal of Sociology*? If it does, it may indicate that the journal carries some authority.

3 Does a professional association publish the journal?

4 Does the journal contain a list of its editorial board and editorial advisers, and are these people well-known in the field?

5 Are articles in the journal refereed, which means do experts review them prior to publication?

5.4.2 Technical publications

There are handbooks and manuals to cater for every area of research interest. Anyone already working in the field of your research would probably be able to tell you what technical publications are available. Equipment manufacturers publish detailed performance specifications (for example those published by motor manufacturers) as do trade associations and bodies such as the British Standards Institution (Sharp *et al*, 2003).

5.4.3 Official publications

The amount of material published on behalf of governments is enormous and forms an important reference source for many types of research (Sharp *et al*, 2003). Government publications are often classified separately in libraries, due to their sheer volume.

Government departments compile and publish statistics, as do many other public bodies. International organisations such as the OECD and UNESCO publish information of international relevance that can be useful for comparison purposes. Many are published in the United Nations Statistical Yearbook.

Denscombe (2003) advises that you should be careful in your use of government statistics. They are an attractive source of information for the researcher, because they are:

- Authoritative, having been produced by the state, with access to huge resources, as well as having expert and professional staff.

- Objective, since they have been produced by officials.

- Factual, as they tend to be the product of unambiguous figures.

However Denscombe (2003) cautions that since government and other official statistics are often produced for the benefit of the government in office, it is possible even for these to be manipulated so that they become a less reliable source of evidence.

5.4.4 Trade association data

Trade and industry associations can provide valuable data about the technical and economic operation of their members. The economic data is more detailed than those found in government statistics.

5.5 The use of secondary data

Secondary information is now **available** in every form and on a **huge scale**. The problem is how to decide what information is required. The use of secondary data will generally come **early** in the process of **marketing research**. In some cases, secondary data may be sufficient in itself, but not always.

Secondary data can:

- Provide a backdrop to primary research
- Act as a substitute for field research
- Be used as a technique in itself

5.5.1 Backdrop to primary research

In **unfamiliar territory**, it is natural that the marketer will carry out some **basic research** in the area, using journals, existing market reports, the press and any contacts with relevant knowledge. Such investigations will aid the marketer by providing guidance on a number of areas.

- Possible data sources
- Methods of data collection (relevant populations, sampling methods)
- The general state of the market (demand, competition and the like)

The often substantial **cost** of primary research **might be avoided** if existing secondary data is sufficient. This data might not be perfect for the needs of the business, though and to judge whether it *is* enough, or whether primary research ought to be undertaken, a cost-benefit analysis should be implemented weighing up the advantages of each method.

There are some situations in which secondary data is bound to be **insufficient**. For instance if your brand new version of an existing product is hugely superior to your competitors' versions because of your unique use of new technology, you have changed the entire market. Primary research will be a necessity to find out the impact of your product.

5.5.3 A technique in itself

Some types of information **can only be acquired** by examining secondary data, in particular **trends over time**. Historical data cannot realistically be replaced by a one-off study and an organisation's internal data would only give a limited picture (Dillon *et al*, 1994).

5.6 Benefits, risks and dangers of reliance on secondary data

Secondary sources of data are often of **limited use** because of the **scope for compounding errors** arising from why and how the data was collected in the first place, who collected it and how long ago.

When considering the quality of the secondary data it is a good idea to consider the following characteristics of it:

(a)   The **producers** of the data (they may have an axe to grind; trade associations may not include data which runs counter to the interest of its members).

(b)   The **reason for the data** being collected in the first place.

(c)   The **collection method** (random samples with a poor response rate are particularly questionable).

(d)   How **old** the data is (government statistics and information based on them are often relatively dated, though information technology has speeded up the process).

(e)   **How parameters were defined**. For instance, the definition of family used by some researchers could be different to that used by others.

(Malhotra, 2004; Dillon et al, 1994)

5.6.1 Advantages and disadvantages of secondary data

Secondary data can be **immensely cost-effective**, but has to be **used with care**.

The **advantages** arising from the use of secondary data include the following.

(a)   Secondary data may solve the problem without the need for any primary research: **time and money is thereby saved**.

(b)   Cost savings can be substantial because secondary data sources are a great deal **cheaper** than those for primary research.

(c)   Secondary data, while not necessarily fulfilling all the needs of the business, can be of great use by:

    (i)   **Setting the parameters**, defining a hypothesis, highlighting variables, in other words, helping to focus on the central problem.

    (ii)   **Providing guidance**, by showing past methods of research, for primary data collection.

    (iii)   **Helping to assimilate the primary research** with past research, highlighting trends and the like.

    (iv)   **Defining sampling parameter**, (target populations, variables).

There are, of course, plenty of **disadvantages** to the use of secondary data.

(a) **Relevance**. The data may not be relevant to the research objectives in terms of the data content itself, classifications used or units of measurement.

(b) **Cost**. Although secondary data is usually cheaper than primary data, some specialist reports can cost large amounts of money. A cost-benefit analysis will determine whether such secondary data should be used or whether primary research would be more economical.

(c) **Availability**. Secondary data may not exist in the specific product or market area.

(d) **Bias**. The secondary data may be biased, depending on who originally carried it out and for what purpose. Attempts should be made to obtain the most original source of the data, to assess it for such bias.

(e) **Accuracy**. The accuracy of the data should be questioned. Here is a possible checklist.

- Was the sample representative?
- Was the questionnaire or other measurement instrument(s) properly constructed?
- Were possible biases in response or in non-response corrected and accounted for?
- Was the data properly analysed using appropriate statistical techniques?
- Was a sufficiently large sample used?
- Does the report include the raw data?
- To what degree were the field-workers supervised?

In addition, was any raw data omitted from the final report, and why?

(f) **Sufficiency**. Even after fulfilling all the above criteria, the secondary data may be insufficient and primary research would therefore be necessary.

The golden rule when using secondary data is **use only meaningful data**. It is obviously sensible to begin with internal sources and a firm with a good management information system should be able to provide a great deal of data. External information should be consulted in order of ease and speed of access starting with directories, catalogues and indexes before books, abstracts and periodicals.

6 Sampling

Sampling is a key topic in all research. Various sampling methods are examined in this chapter.

Definition

A **population** in statistics simply means the set of individuals, items, or data from which a statistical sample is taken. For example you might send a questionnaire to a sample of 100 people who are aged 30 to 40: the population is ALL people aged 30 to 40.

Sampling is one of the most important tools of marketing research because in most practical situations a population will be far too large to carry out a complete survey.

The key to sampling is to remember the practical issues, and its purpose. Once you have decided who you need to invite to participate in research they should be referred to as the '**population of interest**'.

A familiar example of sampling is a poll taken to try to predict the results of an election. It is not practical to ask everyone of voting age how they are going to vote week after week: it would take too long and cost too much. So a sample of voters is taken, and the results from the sample are used to estimate the voting intentions of everyone eligible to vote.

Occasionally a population (set of items) is small enough that **all of it can be examined**: for example, the examination results of one class of students. When the population is examined, the survey is called a **census**. This type of survey is quite rare, however, and usually the researcher has to choose some sort of sample.

A sample is often preferable to researchers who do not have the resources, or the need, to conduct a census. There is often a lot of wastage in conducting a census.

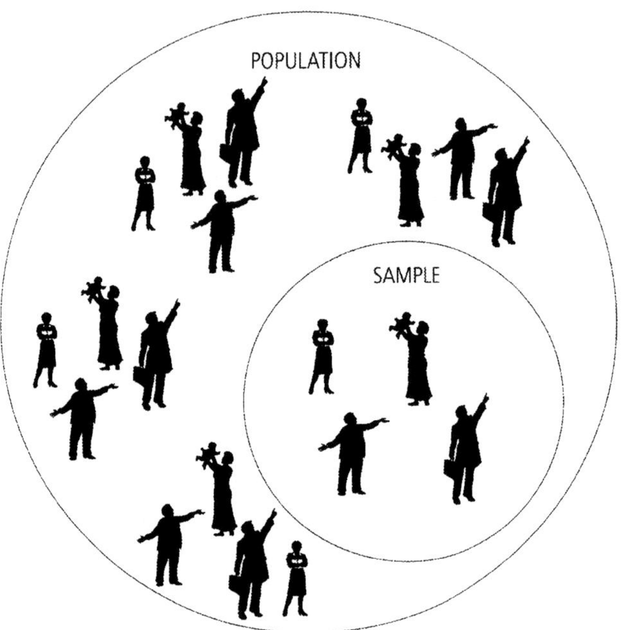

Global case study

Let's think about a practical everyday scenario to demonstrate the benefits of a sample. Imagine you are cooking a pot of pasta. To test whether it was cooked you wouldn't eat the whole pot full and then decide if it was ready because it is likely that you could tell just by testing a small amount. The chances are that if you test a fair proportion of the pasta then you will know if it is cooked or not. This however presupposes that each of the pieces of pasta were sufficiently alike in their characteristics – don't, for example, try cooking linguine and expecting it to cook at the same rate as a whole sheet of lasagne in the same pan.

The same concept is true of people, if there are characteristics which are sufficiently alike within a specific group (population of interest), then it is possible to assume they may have similar beliefs and attitudes, therefore, you will only need to include a proportion of these within your research.

You may think that using a sample is very much a **compromise**, but you should consider the following points.

(a)   It can be shown mathematically that once a certain sample size has been reached, **very little extra accuracy** is gained by examining more items.

(b)   It is possible to **ask more questions** with a sample.

(c)   The **higher cost** of a census may **exceed the value** of results.

(d)   **Things are always changing.** Even if you took a census it could be out of date by the time you completed it.

When designing your sample, you will need to address just five key questions:

1   **Who** do we need to research? – population of interest.

2   **Where** do we find them? – sampling frame.

3   How should we **select** individual respondents? – in other words what sampling technique should be used?

4   **How many** respondents do we need? – this will determine sample size.

5   Will research respondent views be **representative** of the views of everybody else in that situation?

The diagram below outlines the sampling process.

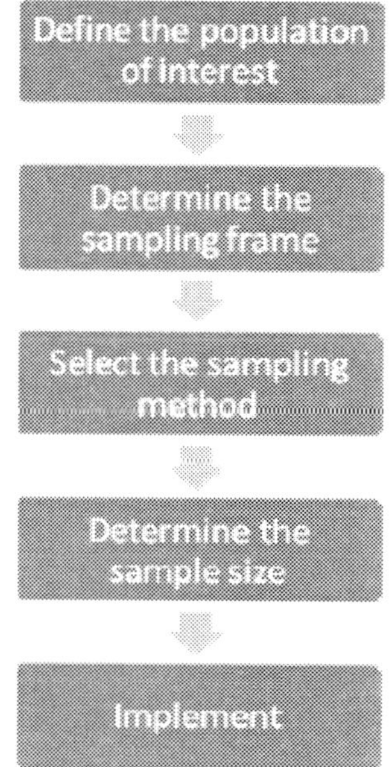

Adapted from Malhotra (2004)

6.1.1 Population of interest

Once you know who should be included within your sample, you should refer to this as your population of interest.

Definition

A **population of interest** refers to the group that the researcher would like to respond within the investigation.

It is essential to correctly define the population of interest eg if the researcher wishes to elicit responses from dedicated gym goers then they should clearly define what constitutes 'dedicated' and plan to only include those individuals in the research.

US research agency Gallup included cell-only (mobile phone only) households in its general population surveys. More than 1 in 8 Americans had ditched land lines in favour of mobiles or internet based phones. There had been concern that the young, those on low incomes and ethnic minority groups had been under represented. A problem for the research agency has been that in the US mobile owners can be charged for incoming calls and so the research agency would need to compensate respondents on top of any incentive given in order to maintain response rates. (Tarren, 2008)

6.1.2 Sampling frames

The next stage is the development of a sampling frame.

Definition

A **sampling frame** refers to the group that the researcher would like to respond within the investigation.

Sampling frames are lists or set of directions for identifying the population of interest.

The term sampling frame originates from science. You may remember your science classes at school when you were given a rectangular frame (similar to an empty picture frame) and were then asked to throw it randomly on the ground. Typical experiments then included counting the number of bugs or plants that you found within that sampling frame.

In market research terms, the picture frame equivalent is usually something tangible from where you can select respondents, such as a database, a directory or instructions about a specific location to visit.

You may have been approached by a survey interviewer while walking in a shopping area (market, retail outlet or mall). The researcher will have been given specific instructions in terms of the time and location from where they should approach shoppers. The directions they would have been given is the sampling frame.

6.1.3 Select the sampling method

The technique you use to select your sample is broadly grouped into one of two types, either a probability sample, which is taken at random, or a non-probability sample, where respondents are selected on characteristics rather than by pure chance.

When **probability samples** are taken, every individual within the population of interest has an **equal chance** of being selected to participate in the research.

When **non-probability samples** are taken the respondents are **not selected by chance** but as a result of planning by the researcher.

The following table shows some of the related qualities of probability versus non-probability sampling.

Research Methods for Global Marketing Practice

Factors affecting sampling approach	Non-probability sampling	Probability sampling
Nature of research	Exploratory	Conclusive
Research flaws likely to be due to sampling or non-sampling errors	Non-sampling errors are larger	Sampling errors are larger
Similarity in population characteristics	Homogenous (similar)	Heterogeneous (different)
Statistical analyses	Not as statistically sound	Statistically sound
Operational issues	Convenient, cost effective, fast	Expensive, time consuming

Adapted from Malhotra (2004)

The diagram below shows the most common types of sample selection methods. You will encounter other methods during your wider reading.

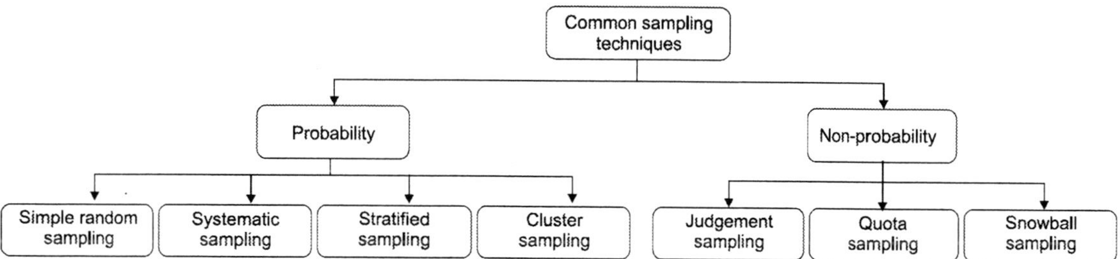

6.2 Probability sampling methods

6.2.1 Random sampling

To ensure that the sample selected is **free from bias**, random sampling must be used. Inferences about the population being sampled can then be made validly.

A simple random sample is a sample selected in such a way that **every item in the population has an equal chance of being included**.

For example, if you wanted to take a random sample of library books, it would **not be good enough to pick them off the shelves, even if you picked them at random**. This is because the **books which were out on loan** would stand no chance of being chosen. You would either have to make sure that all the books were on the shelves before taking your sample, or find some other way of sampling (for example, using the library index cards).

A random sample is **not necessarily a perfect sample**. For example, you might pick what you believe to be a completely random selection of library books, and find that every one of them is a detective thriller. It is a remote possibility, but it could happen. The only way to eliminate the possibility altogether is to take 100% survey (a census) of the books, which, unless it is a tiny library, is impractical.

In many situations it might be **too expensive** to obtain a random sample, in which case quasi-random sampling is necessary, or else it may not be possible to draw up a sampling frame.

6.2.2 Systematic sampling

This technique involves taking your sample at regular intervals from the population, eg to select five people from a bus queue of forty people, you pick every eighth person in the queue.

A small practical problem is selecting a random starting point. This can be done in a variety of ways, such as using random numbers or using the serial numbers on a currency note (but remember to cover yourself by keeping an 'audit trail' [eg photocopy of the page from the random number tables or the pound note/dollar bill etc.] of anything that might be questioned at a later date).

If you have all the items in your sampling frame sequentially numbered on a spreadsheet you can set up a simple formula to identify your sample items, taking into consideration the initial random item.

A potential weakness may arise when using systematic sampling, where there is an element of systematic arrangement of the items in the sampling frame itself. Eg a list of customers names is arranged by name, husband's name followed by wife's name. You may well find that there is an element of systematic arrangement with electoral registers because they tend to show the names by household; father, mother, sons and daughters. However, you will see small families and big families with single person households interspersed in between, effectively breaking up the pattern.

6.2.3 Stratified sampling

This technique involves organising a sampling frame into a number of strata, layers or bands based on specified attributes. Then items are drawn from each strata depending on the relative proportion each strata bears to the total population.

Global case study

You want to select 100 people from a company employing a total 1,500 people made up of 60 managers, 790 sales staff and 650 administration staff. Your stratified sample would be as follows:

	Population numbers	Population percentage	Stratified sample
Managers	60	4%	4
Sales staff	790	52.7%	53
Admin staff	650	43.3%	43
Total	1,500	100%	100

This technique provides some control over the representativness of the sample. In the above example, with only 60 managers, a random selection could have caused a bias in the number of managers selected if say 2 or 6 of them were selected causing under- or over-representation.

Having determined the number of items to be selected from each strata, you can then apply random sampling or systematic sampling to pick the individual items.

Various attributes can be used to stratify your population. Examples include:

- Salary bands
- Age bands
- Department
- Geographic region
- Profession or trade
- Gender

You may also wish to use more than one attribute to drive your classification, eg by salary band per department.

6.2.4 Cluster sampling

This technique involves dividing the population into a number of convenient clusters and randomly selecting a certain number of clusters for research. Hence your sampling frame would not comprise individual people or items but groups of people or items.

Research Methods for Global Marketing Practice

In a survey of petrol station staff training levels in Ghana, instead of selecting a sample randomly from petrol stations across the country, the county is broken down into regions and a random selection taken of such areas eg Greater Accra, Ashanti and Volta etc. Then all the petrol stations in the selected areas would be subjected to the specified research enquiries regarding their staff training levels.

This technique assumes that the composition of individual clusters are similar to each other, though the contents of individual clusters may not be homogeneous. Various approaches can be used to form the basis of clustering.

- Geographic areas
- Companies or businesses
- Offices within an organisation

Beware of trying to cluster on the basis of departments within an organisation, eg the opinions of people in a production department may well be materially different to those of people comprising the marketing department.

6.3 Non-probability sampling

Non-random sampling is used **when a sampling frame cannot be established**.

6.3.1 Judgemental sampling

Judgemental sampling involves selecting respondents because they possess particular characteristics which the researcher believes are **representative of the population of interest** as a whole. 'Typical' residents for example may be selected from a selection of streets that a researcher believes are representative of an entire neighbourhood.

Snowball sampling (see below) is sometimes considered to be a form of judgemental sampling (Dillon *et al*, 1994) because this method involves the respondent suggesting other individuals for selection because they are similar to themselves. Snowball sampling is used when respondents are hard to find and tends to be only relevant for small samples in qualitative studies.

6.3.2 Convenience sampling

Convenience sampling refers to samples that are selected because the **population of interest is easy to access** by the researcher. Convenience samples are sometimes considered to be biased and unprofessional (Wilson, 2006) however if the easiest to access respondents are reasonably similar to the population of interest then it can be a justifiable method on a cost and resource basis.

Many online surveys use convenience sampling because it is difficult to establish common patterns between users of websites (Wilson, 2006).

6.3.3 Quota sampling

In quota sampling randomness is forfeited in the interests of **cheapness and administrative simplicity** Investigators are told to interview all the people they meet up to a certain quota. A large degree of bias could be introduced accidentally. For example, an interviewer in a shopping centre may fill his quota by only meeting people who can go shopping during the week. In practice, this problem can be **partly overcome by subdividing the quota** into different types of people, for example on the basis of age, gender and income, to ensure that the sample mirrors the structure or stratification of the population. The interviewer is then told to interview, for example, 30 males between the ages of 30 and 40 from social class B. The actual choice of the individuals to be interviewed, within the limits of the **quota controls**, is left to the field worker.

6.3.4 Snowball sampling

With snowball sampling, you would use a respondent to help you identify your next respondent. The technique is usually associated with phenomenological research and the samples obtained might not be representative.

You can of course begin with more than one respondent and each respondent can be asked to steer you to more than one other respondent.

The advantage is that the *multiplier effect* can give you a reasonable sized sample.

The technique is compatible with purposive sampling, eg a respondent can be asked to identify the next person with certain attributes in terms of gender, age, organisational status etc. This method is very effective when it is not easy to identify respondents from a sampling frame.

Global case study

A researcher wanted to investigate reasons for shoplifting. Clearly it is not likely that they would gain access due to privacy issues to a ready made list of thieves from which to draw a sample. The researcher managed to identify one respondent and from there managed to generate a substantial sized sample from their known accomplices. Given their close proximity, the findings are not likely to be representative of all shoplifters but will be useful for that region.

6.4 The size of a sample

As well as deciding on the appropriateness of a particular sampling method for a given situation, the size of the sample actually selected must also be given consideration.

There are three key factors that influence sample size.

- Population size
- Confidence sought
- Precision of results

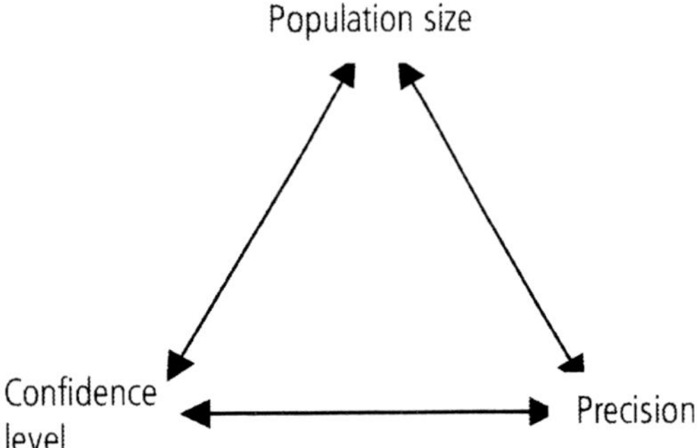

Population. Sample size is a function of population size. The larger the population, the bigger the sample you will require. However, the relationship is not altogether directly proportional, as can be seen from the table below.

Confidence. This is the level of certainty that the data collected reflects the characteristics in the population. It is usually expressed as a percentage, eg 95%, 97% etc, though in practice, 95% is usually sought.

Precision. This refers to how accurately you will be able to express your result. In market research for example, this is referred to as *margin of error*.

The above principles can be seen from a review of the following table.

Research Methods for Global Marketing Practice

Population	Margin of error			
	5%	3%	2%	1%
50	44	48	49	50
100	79	91	96	99
150	108	132	141	148
200	132	168	185	196
250	151	203	226	244
300	168	234	267	291
400	196	291	434	384
500	217	340	414	475
750	254	440	571	696
1,000	278	516	706	906
2,000	322	696	1,091	1,655
5,000	357	879	1,622	3,288
10,000	370	964	1,936	4,899
100,000	383	1,056	2,345	8,762
1,000,000	384	1,066	2,395	9,513
10,000,000	384	1,067	2,400	9,595

Global case study

Your business has 1,000 employees, some of whom wear spectacles and other who do not. You wish to use sampling to establish the breakdown between spectacle wearers and non-spectacle wearers.

You want to be 95% confident that your research results are within plus or minus 3% (precision) of the real breakdown in the total population.

From table you see that a sample of 516 will be required. You use random sampling to identify employees and your sample shows the following results.

	Number	Sample percentage	You can be 95% confident that the situation in the population is in the range between
Spectacle wearers	148	29%	26% and 32%
Spectacle non-wearers	368	71%	68% and 74%
Total	516	100%	

If you wanted a more precise result of say a 1% margin of error you would, from the table, have to use a sample of 906.

The following chart shows the sample sizes required for populations of 1,000, 10,000, 100,000, 1,000,000 and 10,000,000 at 95% confidence and 3% precision.

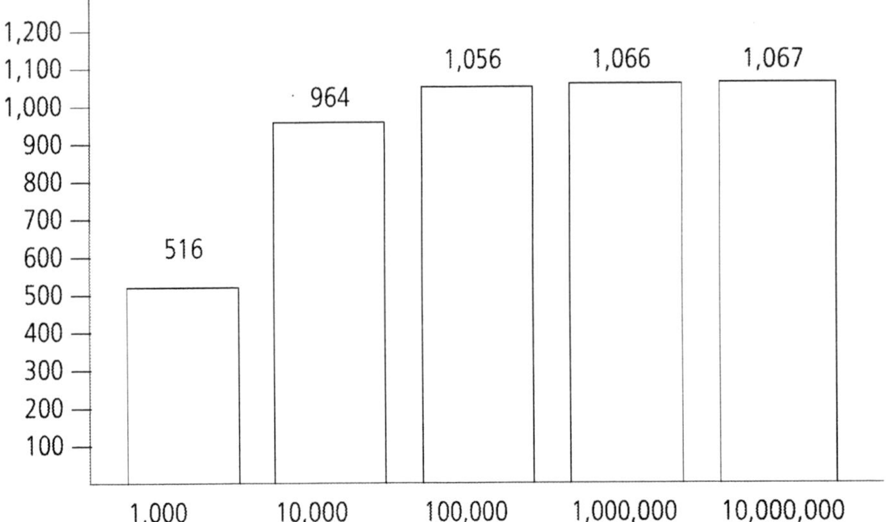

Although, in many circumstances, statistical processes should be used to calculate sample sizes, as we'll see in a moment there is **no universal law** for determining the size of the sample. Researchers may simply rely on their **experience** from other studies similar to the project in hand. Two general considerations should, however, be borne in mind.

(a)   The larger the size of the sample, the more accurate the results.

(b)   There reaches a point after which there is little to be gained from increasing the size of the sample.

Despite these principles other, more administration-type factors, play a role in determining sample size.

(a)   **Money** and **time** available.

(b)   **Degree of precision required.** A survey may have the aim of discovering residents' reaction to a road widening scheme and hence a fairly small sample, producing imprecise results, would be acceptable. An enquiry into the safety of a new drug would, on the other hand, require an extremely large sample so that the information gained was as precise as possible.

(c)   **Number of subsamples required.** If a complicated sampling method such as stratified sampling is to be used, the overall sample size will need to be large so as to ensure adequate representation of each subgroup (in this case, each stratum).

(d)   **Blind guess** (although this is not really acceptable in most circumstances).

(e)   **Industry standards** or rules of thumb; certain industries assume particular benchmark sample sizes eg 200-300 for a product test.

If you type into an internet search engine a term like 'sample size calculator', you will find that there are a number of websites where you can download, for free, a calculator to help you work out statistically your sample size.

6.5 Problems with sample data and research designs

There are many **potential problems** with sample data including **bias, unrepresentative data,** and **insufficient data,** perhaps because of non-response.

There are several faults or weaknesses which might occur in the design or collection of sample data. These are as follows.

Research Methods for Global Marketing Practice

(a) **Bias**. In choosing a sample, unless the method used to select the sample is the random sampling method, or a quasi-random sampling method, there will be a likelihood that some 'units' (individuals or households etc) will have a poor, or even zero chance of being selected for the sample. Where this occurs, samples are said to be biased. A biased sample may occur in the following situations.

    (i) The sampling frame is out of date, and excludes a number of individuals or 'units' new to the population.

    (ii) Some individuals selected for the sample decline to respond. If a questionnaire is sent to 1,000 households, but only 600 reply, the failure of the other 400 to reply will make the sample of 600 replies inevitably biased.

    (iii) A questionnaire contains leading questions, or a personal interviewer tries to get respondents to answer questions in a particular way.

(b) **Insufficient data**. The sample may be too small to be reliable as a source of information about an entire population.

(c) **Unrepresentative data**. Data collected might be unrepresentative of normal conditions. For example, if an employee is asked to teach a trainee how to do a particular job, data concerning the employee's output and productivity during the time he is acting as trainer will not be representative of his normal output and productivity.

(d) **Omission of an important factor**. Data might be incomplete because an important item has been omitted in the design of the 'questions'.

(e) **Carelessness**. Data might be provided without any due care and attention. An investigator might also be careless in the way he gathers data.

(f) **Confusion of cause and effect (or association)**. It may be tempting to assume that if two variables appear to be related, one variable is the cause of the other. Variables may be associated but it is not necessarily true that one causes the other.

(g) Where questions call for something **more than simple 'one-word' replies**, there may be difficulty in interpreting the results correctly. This is especially true of 'depth interviews' which try to determine the reasons for human behaviour.

One method of checking the accuracy of replies is to insert control questions in the questionnaire, so that the reply to one question should be compatible with the reply to another. If they are not, the value of the interviewee's responses are dubious, and may be ignored. On the other hand, the information that the interviewee is genuinely confused about something, and so offers contradictory answers, may be valuable information itself, or it may reflect the way the questions are structured.

6.5.1 Non-sampling error

A non-sampling error is an error that results solely from the manner in which the observations are made, and leads to inaccurate conclusions being drawn from the group being studied. In other words, there is a problem with the way the data is collected. It can occur whether a total population or a sample is being used. The simplest example of a non-sampling error is inaccurate measurements due to poor procedures or data input errors. Unintended errors may result from any of the following.

- The manner in which the response is elicited – no two interviewers are alike, and questions may be worded poorly

- The suitability of the persons surveyed – some may give deliberately inaccurate answers

- The purpose of the study – if the respondent knows what it is, it may affect the responses given

- The personal biases of the interviewer or survey writer – questionnaires must be designed to draw out useful responses

- Non-response – either through refusal or non-availability

6.5.2 Non-response

Non-response (of a sample member) cannot be avoided. It can, however, (apart from in mail surveys) be kept at a reasonable level. Experience has shown that the non-response part of a survey often differs considerably from the rest. The types of non-response are as follows.

(a)   **Units outside the population**. Where the field investigation shows that units no longer exist (eg demolished houses), these units should be considered as outside the population and should be subtracted from the sample size before calculating the non-response rate.

(b)   **Unsuitable for interview**. This is where people who should be interviewed are too infirm or too unfamiliar with the language to be interviewed.

(c)   **Movers**. People who have changed address since the list was drawn up cannot be interviewed.

(d)   **Refusals**. Some people refuse to co-operate.

(e)   **Away from home**. People might be away from home for longer than the field work period and call-back might not be possible.

(f)   **Out at time of call**.

These sort of problems occur chiefly in **random** sample surveys. Some of the above do not apply when interviewing is done in factories, colleges or offices. In quota sampling (c), (e) and (f) do not appear. Although the interviewer may miss some people for these reasons, he or she simply continues until he or she fills the quota.

Social change can influence the level of non-response. Rising crime means that householders may be afraid to answer the door to strangers and there are other employment opportunities for 'doorstep interviewers'. Response rates are therefore slipping as more people either refuse to be or cannot be interviewed.

Global case study

The director general of the Market Research Society, the industry's professional body, says falling response levels are not just a UK phenomenon. He is concerned that the quality of research will begin to be affected, for the lower response rates are, the greater the departure from ideal cross-sections of opinion, and the less accurate findings are likely to be.

Another problem is that of **'data fatigue'**, as the public becomes tired of filling in questionnaires and more cynical about the real motives of 'market researchers' because of 'sugging' (selling under the guise of research) and 'frugging' (fundraising under the guise of research).

6.5.3 How to deal with non-response

Taking **substitutes** (such as the next house along) is no answer because the substitutes may differ from the non-respondents. Instead the interviewer can try to increase the response rate.

(a)   Little can be done about **people not suitable** for interview.

(b)   **People who have moved** are a special category. It is usually not practical to track them down. It is acceptable to select an individual from the new household against some rigorously defined procedure.

(c)   To minimise **'refusals'**, keep questionnaires as brief as possible, use financial incentives, and highly skilled interviewers. Refusal rates tend to be low (3-5 per cent).

(d)   People **'away from home'** may be contacted later, if this is possible.

(e)   People **'out at time of call'** is a common problem. The researcher should plan the calling time sensibly (for example, as most people employed are out at work in the day-time, call in the evening). Try to establish a good time to call back – or arrange an appointment.

7   Validity, reliability and generalisability

We briefly discussed the concepts of validity and reliability in Chapter 1 but need to review the concept here because it is an important consideration when designing your research strategy.

> **Definition**
>
> **Validity** refers to the extent that the research method or instrument actually measures or records what the research set out to measure or record.
>
> **Reliability** refers to whether the research could be replicated if repeated with the same results because the measure is consistent and stable.
>
> **Generalisability** refers to the extent to which the findings of the research are more generally applicable outside of the specifics of the research.
>
> (Robson, 2009)

Validity and reliability are important in order to persuade audiences (and yourself!) that the findings of your research are trustworthy. Although they are separate concepts according to Bryman and Bell (2007) validity presumes reliability and so if your measure is not reliable it cannot be valid.

Global case study

When people argue that IQ scores do not really reflect their level of intelligence, they are questioning the measurement validity of the IQ test.

You may have noticed from the example above that many of these points are most relevant when we look at designing specific data collection tool. Frequently there are debates about whether constructs have been accurately measured, how for example can the constructs of satisfaction and loyalty be measured when there are often many contextual differences and differences between people. There is also however a requirement at the overall research design stage to consider whether there are any factors in the overall design that may impede validity and reliability.

Global case study

Research may be conducted amongst airline cabin crew to identify areas of service they considered beneficial to adapt to improve customer service levels. At the time of the research project however they were in the process of taking industrial action against their employer and therefore their views may be biased in relation to this issue which would undoubtedly impact their own job roles. The internal validity of the research will therefore be compromised.

We will return to these concepts in Chapters 4 and 5.

8 The ethical elements of research

8.1 Definition

> **Definition**
>
> Remenyi *et al* (1998) define **ethics** as a sense or understanding of what is right or wrong.

8.2 Ethical issues

In the world of scientific research, there are many obvious ethical issues that need to be addressed, for example:

- How far should research on human cloning go?
- To what extent should the scientist explore the development of weapons of mass destruction?
- Should scientific experiments that cause pain and suffering to live animals be carried out?

Although these are issues of serious concern, the implications of ethical considerations in management research, are no less important because this is your chosen sphere of work. Generally, management research is often involved with human participants and it is important to ensure that they are properly protected.

Remenyi *et al* (1998) acknowledge that ethical issues in business research will indeed be less 'lofty' than those of scientific research but does give some examples of areas of concern in management research.

- Implications of employment of new technologies on those employed in the production function in a factory

- Marketing practices that might impinge on privacy, or which might exert excessive influence or stress on prospective customers

- Ways of controlling or manipulating a workforce

Remenyi *et al* (1998) suggest that although there are ethical concerns regarding *what* is researched, perhaps of greater concern are questions of *how* the research should be conducted.

8.2.1 Ethical context

Saunders *et al* (2009) suggest that '*in the context of research, ethics refers to the appropriateness of your behaviour in relation to the rights of those who become the subject of your work, or are affected by it*'.

They also suggest that '*the appropriateness or acceptability of our behaviour as researchers will be affected by broader social norms of behaviour*'.

Anglia Ruskin University expects the ultimate ethical standards to be observed in the conduct of your work. The university's standards are driven by the imperatives of *nonmaleficience* (do no harm) and *beneficence* (do good):

- The potential benefits of the research to the participants, the scientific (business) community and/or society must be clearly stated

- Procedures must be justified, explaining why alternative procedures involving less risk cannot be used

- Any cultural, religious, gender or other differences in a research population should be sensitively and appropriately handled by researchers at all stages

Research Methods for Global Marketing Practice

8.2.2 Courtesy

Collis and Hussey (2003) emphasise matters they refer to as *common courtesy*. These include

- Thanking individuals and organisations for their assistance with your research, both verbally at the time and afterwards by letter

- Keeping your promises to provide copies of transcripts of interviews or the final report/dissertation

Being courteous to people you come into contact with during your research is more likely than not to help you complete your dissertation smoothly and efficiently.

8.2.3 Researcher's reputation

An aspect which should not be forgotten is your own reputation, which as a professional person should be preciously guarded. A good name, once damaged, will be difficult to repair.

On the practical side, if your research is being carried out in a particular industry it might be beneficial from a career enhancing viewpoint to ensure that you are well thought of and respected in that industry.

8.3 Power in organisations

8.3.1 Power and structure

In most organisations, such as businesses or universities, there is likely to be a formal as well as an informal power structure. Some people will hold formal positions of power whereas other people will exercise significant influence. For example, the managing director of a company may hold the top position in a company, but his or her personal assistant may exert lots of influence. It is sensible to develop an early understanding of the people within any organisation who have the power or influence to help you complete your dissertation successfully.

8.3.2 Management expectations

You may be conducting your research within your own organisation or you may be an external researcher in another organisation. In either case, you will have been granted access to information by the management of the company. This means that they have done you a favour (Bell, 2005) and you are therefore beholden to them, or even in their debt.

In return for allowing you access, management of the company may expect an uncritical analysis to be painted, or they may expect unlimited access to the opinions you glean from the staff about the company. This inferred existence of power exercised by management of the company over you can present you with an ethical dilemma, as it may cause bias to enter into your work.

8.3.3 Example of use of power

Easterby-Smith *et al* (2002) provide an example that might illustrate how the assumption of power of a researcher over a research subject can sometimes be turned upside-down.

Global case study

A senior academic interviewed a member of the British Royal Family. At the end of the interview the interviewer offered to have the tape transcribed and to send the Royal personage the transcript for him to delete anything which he did not like. The Royal personage stretched out a hand and said, *'No, I shall retain the tape and I shall let you have the portions I am prepared to have published.'*

(Easterby-Smith *et al*, 2002)

8.4 Exercise of ethical standards

8.4.1 Areas of concern

There is broad consensus among authors regarding the areas where it is important to exercise ethical standards.

- Informed consent of research participants
- Potential deception of participants
- Confidentiality and anonymity
- Privacy and dignity
- Freedom from coercion
- Integrity of evidence

8.4.2 Informed consent

Definition

Marshall (1997) suggests that **informed consent** means that researchers have satisfied themselves that the participants had an adequate understanding of the risks they would incur, before they gave their consent. They explain that the term 'adequate' understanding is used because for practical purposes, full understanding will not be attainable.

Researcher obligations

Denscombe (2003) argues that *'there is some moral obligation on the researcher to protect the interests of those who supply information and to give them sufficient information about the nature of the research so that they can make an informed judgement about whether they wish to co-operate with the research'*.

The following matters should be made clear to participants:

- Purpose and nature of the research
- The implications of participation
- Benefits intended to result from the research
- The risks, including psychological distress

8.4.3 Potential deception of participants

Scope for deception

Saunders *et al* (2009) suggest that although research subjects have agreed to participate in research, they may nevertheless still be subject to deception.

- Researchers might misrepresent the nature of the research to the research subjects. This is most likely to occur where experiments are being performed (Sekaran, 2009).

- There may be concealment or lack of full disclosure of the purpose and sponsorship of the research (Zikmund, 2002).

- Another associated organisation might use the data collected for its commercial advantage.

- The use of *covert observation* where access is denied or the researcher considers that full disclosure of the purpose of the research would unduly influence the research subjects and therefore distort the research findings. The technical term for this phenomenon is known as the research subject's 'reactivity'.

Participant observation

According to Ditton (1977) the use of (covert) participant observation research methods are essentially deceitful. The researcher participates in a situation where he or she is at the same time observing and recording (perhaps later) what has taken place, without the participants knowing.

The view of Denscombe (2003) reinforces this by suggesting that *'when researchers opt to conduct full participation, keeping their true identify and purpose secret from others in the setting, there are ethical problems arising from the absence of consent on the part of those being observed, and of deception by the researcher'*.

8.4.4 Confidentiality and anonymity

Participants' rights

Saunders *et al* (2009) refer to the participants' right to expect agreed anonymity and confidentiality to be strictly observed in relation to discussions with other research or organisational participants and during the reporting of findings (including from those who gain subsequent access to data). They advise that *'once promises about confidentiality and anonymity had been given, it is of great importance to make sure that these are maintained'*.

Easterby-Smith *et al* (2002) advise that researchers must exercise due ethical responsibility by not publicising or circulating any information that is likely to harm the interests of individual informants, especially the less powerful ones.

Non-sale of respondent details

Zikmund (2002) warns that organisations may be *'keen to exploit the business opportunities that might potentially exist in any of your survey respondents'*. He cautions that it is *'the researcher's responsibility to ensure that the privacy and anonymity of the respondents is preserved'*. You should not pass on the personal details of your respondents, without their permission, to your sponsoring organisation. Nor should you attempt to sell any of this information.

Data protection regulations

Consideration should also be taken of any data protection legislation in your local environment.

8.4.5 Privacy and dignity

Cultural influences

Privacy is a concept that is probably subject to strong cultural influences and hence a researcher will need to be aware of the interpersonal boundaries of the locality or society in which the research is conducted.

Global case study

Westerners with an individualist culture who visit Far Eastern countries, that in general terms have a group-oriented culture, find it awkward to deal with routine questions by local people they meet concerning personal details such as age, marital status, children, occupation, educational achievement, income and address.

Whereas in the west privacy is prized, in the east openness is essential to foster the trust and acceptance implicit in a group culture.

While these may reflect stereotypes of the *dominant* social paradigms of west and east, they may be said to have *back-up* social paradigms, eg western individualism is complemented by teamwork whereas eastern group orientation operates alongside the importance of personal independence.

Embarrassment and ridicule

Collis and Hussey (2003) argue that in research it would not be ethical to embarrass or ridicule participants, but unfortunately this can easily be done. The relationships between the researcher and the phenomenon being studied is often complex.

Activity 1

Consider whether a researcher would be respecting an unsuspecting research subject's privacy by viewing video recordings taken by hidden cameras viewing the following places:

(a)   The bakery isle of a supermarket

(b)   The personal toiletries sections in a chain of chemist shops

Care should be exercised not only in what a respondent is asked but also what a respondent is asked to do, listen to, watch or taste.

Global case study

People may have cultural, religious or humanitarian reasons for not eating or drinking various products such as beef, pork, meat in general, shellfish and alcohol. Care should be taken not to place a potential respondent in an embarrassing position of being asked questions about something which might be offensive. This is particularly relevant if your research includes cross-cultural investigations.

Dignity

Dignity refers more to the research subject's right not to be embarrassed or humiliated in any way. In practice, this involves not only the questions a participant is asked or any actions they are required to perform, but also the way they are treated and spoken to.

Global case study

You may need to interview the poorest performing employee in a company, department or group to find out his or her views on their situation. Such a person is probably in a psychologically vulnerable state of mind and will need to be handled with tact, diplomacy and sensitivity.

Activity 2

Your company manufactures and markets a recently launched brand of deodorant. They have agreed to sponsor research on consumer buying behaviour in the market for deodorants.

Your research is to include interviewing a sample of consumers. What types of questions should you be careful to avoid on grounds of privacy and dignity?

Sekaran (2009) counsels that '*personal or seemingly intrusive information should not be solicited, and if it is absolutely necessary for the project, it should be tapped with high sensitivity to the respondent, offering specific reasons therefore*'.

Activity 3

It might be a good idea at this juncture to consider areas where potential participants' rights to privacy should be guarded. Jot down areas which you would avoid asking questions about during your research.

Saunders *et al* (2009) suggest that '**privacy may also be affected by the nature and timing of any approach which you make to intended participants, say by telephoning at "unsociable" times, or, where possible, by "confronting" intended participants**'. Zikmund (2002) describes this aspect in terms of '*Is the telephone call that interrupts someone's favourite television programme an invasion of privacy?*'

Culturally there are likely to be global differences between what may or not be an invasion of privacy, the following are outlined as potential areas to consider within the context of your own research.

Personal

- At meal times

- When the potential respondent is busy doing something

- When the potential respondent is not in a state to see or talk to you, eg

    - just got out of bed
    - just about to receive visitors
    - home is being redecorated

- When the potential respondent is about to go out

Organisational

- In busy periods
- When the people involved are in a meeting
- Generally when they have other priorities

8.4.6 Integrity of evidence

Objectivity

Zikmund (2002) emphasises the importance of ensuring accuracy via objectivity. He outlines some examples of what the ethical researcher should consider.

- Selective use of data to prove a particular point
- Misrepresenting the statistical accuracy of your data
- Overstating the significance of the results by altering the findings

Remenyi *et al* (1998) reinforce the above message in warning that any attempt to '*window dress or manipulate and thus distort is of course unethical, as is any attempt to omit any inconvenient evidence*'.

Zikmund (2002) explains that it is assumed that the researcher has the obligation to analyse the data honestly and report correctly the actual data collection methods. He warns that '*hiding errors and variations from proper procedures tends to distort or shade the results. A more blatant breach of the researcher's responsibilities would be the outright distortion of data*'.

Remenyi *et al* (1998) provide some sound advice and perhaps a modicum of reassurance regarding the processing of research evidence. They advise that it is *'not a useful or rational strategy to fabricate evidence or deliberately to misrepresent it, as a master or doctoral degree does not rely on the candidate finding or proving a particular result. Even where hypotheses or theoretical conjectures are rejected, the research is perfectly valid and there is no reason why such findings should not lead to the awarding of the degree'.*

Global case study

Sir Cyril Burt was a famous British psychologist whose work on the relationship between IQ and genetic factors led for a time to IQ testing in British schools.

However, after his death doubts surfaced about the statistical reliability of his work and even the existence of his two research assistants.

Unfortunately, he is alleged to have placed his papers into six chests which all got burned.

(Saunders *et al*, 2009)

Retention of research records

The above example highlights the importance of keeping proper records of your research work. According to Remenyi *et al* (1998) *'the original source of evidence, for example, a transcript of an interview, or copies of the original questionnaire, should be kept for a period of time, say somewhere between two and five years, to allow other researchers to access the data'.*

Replicability

Marshall (1997) suggests that it is *'incumbent on researchers, therefore, to do all they can to facilitate replication of their work by others'.*

- This requires full disclosure of methodology used.
- Findings should be presented in quantitative form, rather than prosaic form.
- The language used should be unambiguous.

Global case study

Medically it is difficult to graft skin from one person to another because of the problem of rejection, unless certain unusual circumstances prevail, eg the donor and recipient are identical twins.

In 1974, a Dr William Summerlin, of the Sloan-Kettering Institute of New York announced that he had discovered a 'tissue culture' technique that would lead to successful skin transplants.

He had trouble in convincing the scientific community regarding his technique so he produced two white mice with black patches, which were purported to have been transplanted from a black donor mouse. What Summerlin did was to use a black felt tip pen to touch-up the grafts he had made to the white mice.

Scientists were not able to *replicate* the results and eventually the fabrication was uncovered.

Saunders *et al* (2009) emphasises that the duty of researchers to represent their data honestly extends to the reporting stage of a dissertation and the importance of sustaining objectivity.

Remenyi *et al* (1998) reinforce this message and stress the importance of ensuring the results are not produced in such a way as simply to support the opinions or prejudices of the researcher. They suggest that sometimes personal bias is so subtle that it might unconsciously creep into the presentation of findings. They argue an approach for dealing with this is to articulate the bias and leave it to the readers to compensate for it. They cite Gould (1980) when he said:

'*Science (as well as management) is not an objective, truth-directed machine, but a quintessentially human activity, affected by passion, hopes and cultural bias.*'

1 Appreciate the purpose of a research proposal

- To establish the feasibility of your research.

- To be used as a working document to develop as ideas are refined during the research process.

- To enable you to progress to the Research Project for Global Marketing Practice.

2 Identify components of a research proposal

- Background and academic context, literature review, research methods, timescale and resources.

3 Evaluate types of data and their associated properties

- Qualitative research generates contextually rich data.

- Quantitative research provides quantifiable data.

- Primary data is collected specifically for the purposes of the investigation.

- Secondary data already exists and has been collected for another purpose.

4 Evaluate the value of mixed methods approaches to research

- Combines primary and secondary data collected in a variety of methods.

- Useful in order to cross-check findings through triangulation.

5 Critically review the value of secondary data

- Advantages include cost and time savings, sometimes of high quality depending on the source.

- Disadvantages include that it does not necessarily address the research questions, may be old, reliability needs careful assessment and it can be difficult to access.

6 Apply sampling procedures within your research project

- Samples are used when it is not possible to include the entire population of interest within your research project eg conduct a census.

- The sampling process includes identifying: who/what we are going to include (population of interest), where they can be found (sampling frame), how should they be selected (sampling method) and how many should be included (sample size).

- Probability sampling methods mean that every person/object within the population has an equal chance of selection. The methods include random, systematic, stratified and cluster sampling.

- Non-probability methods mean that respondents are not selected by chance but by planning by the researcher. Methods include judgemental, quota and snowball sampling.

7 Recognise the importance of validity and reliability in research

- Validity refers to the research measuring what it intended to measure.

- Reliability refers to the measure being valid and so if the research were to be repeated the same result would be obtained.

8 Apply ethical principles in your research

- Ethical practice must be maintained in terms of the topic you are researching and how you conduct your research.

1   Response to hidden cameras:

(a)   This is probably quite acceptable because a person's selection of bread is not likely to involve privacy issues.

(b)   This is probably unethical because there are many toiletries that are of a very personal nature to either men or women, which they would prefer to select unobserved.

2   You might explore what they think about the product itself, its feel, its scent, its effectiveness, how it compares with other deodorants used, etc. However, you might perhaps avoid asking how many times they use it a day, whether they have a body odour problem, how often they bathe or shower etc.

3   Sensitive areas might be:

- Age
- State of health/medical history
- Race
- Family background
- Income
- Sexual orientation
- Family/relational status

Bell, J., (2005). *Doing Your Research Project: A Guide for First Time Researchers in Education, Health and Social Science.* 4th edition. Buckingham: Open University Press.

Bryman, A. and Bell, E., (2007). Business Research Methods. Second edition. Oxford: Oxford University Press.

Cohen, L. and Manion, L., (1994). *Research Methods in Education.* 4th edition. London: Routledge.

Collis, J. and Hussey, R., (2003). *Business Research: A Practical Guide for Undergraduate and Postgraduate Students.* 2nd edition. Basingstoke: Macmillan Palgrave.

Denscombe, M., (2003). *The Good Research Guide for Small-Scale Social Research Projects.* 2nd edition. Buckingham: Open University Press.

Dillon, W., Madden, T. & Firtle, N., (1994). *Marketing Research in a Marketing Environment.* 3rd edition. Illinois: Irwin.

Ditton, J., (1977). *Part-time crime: an Ethnography of Fiddling and Pilferage.* London: Macmillan.

Easterby-Smith, M., Thorpe, R. and Lowe, A., (2002). *Management Research: An Introduction.* 2nd edition. London: Sage.

Malhotra, N., (2004). *Marketing Research: an applied orientation.* New Jersey: Pearson

Marshall, P., (1997). *Research Methods How to design and conduct a successful project.* Oxford: How to Books. Transatlantic publications.

Remenyi, D., Williams, B., Money, A. and Swartz, E., (1998). *Doing Research in Business and Management: An Introduction to Process and Method.* London: Sage.

Robson, C., (2009). *Read World Research: A Resource for Social Scientists and Practitioner Researchers.* Oxford: Blackwell.

Saunders, M., Lewis, P. and Thornhill, A., (2009). *Research Methods for Business Students* (5th ed.) Harlow: Pearson Education Limited.

Sekaran, U., (2009). *Research Methods for Business: A Skills Building Approach.* New York: Wiley.

Sharp, J. A., Peters, J. and Howard, K., (2003). *The Management of a Student Research Project.* 3rd edition. Milton Keynes: The Open University.

Stewart, D.W. and Kamins, M.A., (1993). *Secondary Research: Information Sources and Methods.* 3rd edition. Newbury Park: Sage.

Tarren, B., (2008). 'Pollster Gallup adds cell-only homes to general population sample' *Research*, London: MRS, February 2008 Edition, p. 6.

Zikmund, W.G., (2002). *Business Research Methods* (7th ed.) South-Western College Publishers

Wilson, A., (2006). Marketing Research An Integrated Approach. 2nd edition. Harlow: Prentice Hall.

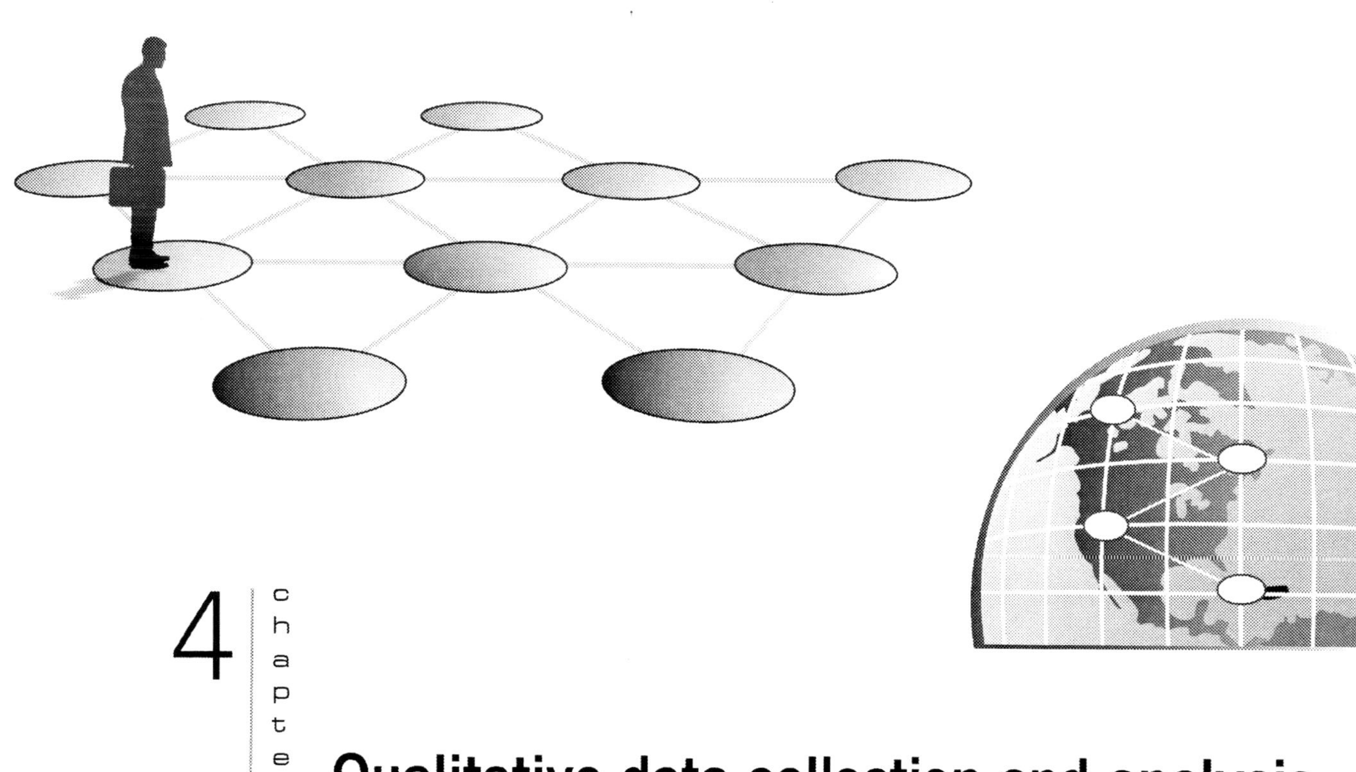

4 chapter

Qualitative data collection and analysis

Qualitative research is a process which aims to collect primary data. Its main methods are the open-ended interview, whether this be a **depth interview** (one-to-one) or a **group discussion** (focus group), and **projective techniques**. The form of the data collected is narrative, rather than isolated statements which can be quantifed. Pure qualitative researchers would argue that there should never be an attempt to count or assign numerical values to qualitative data. The main purpose is to **understand** consumer behaviour and perceptions rather than to measure them.

Qualitative research is 'research which is undertaken using an unstructured research approach with a small number of carefully selected individuals to produce non-quantifiable insights into behaviour, motivation and attitudes.' (Wilson, 2006).

We'll begin this chapter by considering when it might be **appropriate** to conduct qualitative research. Then we go on to consider each of the main methods before considering how qualitative research can be analysed.

Contents

By the end of this chapter you will be able to:

- Describe the nature of qualitative data
- Evaluate the use of alternative data collection tools in your research project
- Analyse qualitative data

1 The nature of qualitative data

Qualitative research may be conducted in dozens of ways (Miles and Huberman, 2003) and even entire texts devoted to qualitative methods rarely cover them all.

In the last 25 years or so there has been an incredible growth within academic research and practice based marketing research (Miles and Huberman, 2003) in the use of qualitative research.

You will remember from the previous chapter that qualitative data tends to be wordy, not numbers orientated and is contextually rich in descriptions and explanations. The real-life stories and incidents described within qualitative studies provide vivid and meaningful insight into the research topic.

Miles and Huberman (2003) argued that when organised, the vivid meaningful flavour of qualitative findings are more convincing to the reader (another researcher, policy maker or practitioner) than pages of summarised numbers (eg the output from quantitative research).

Although as we have already stated there are many qualitative data collection tools, the most common methods of qualitative data collection used in practice (within a marketing context) are shown in the diagram below.

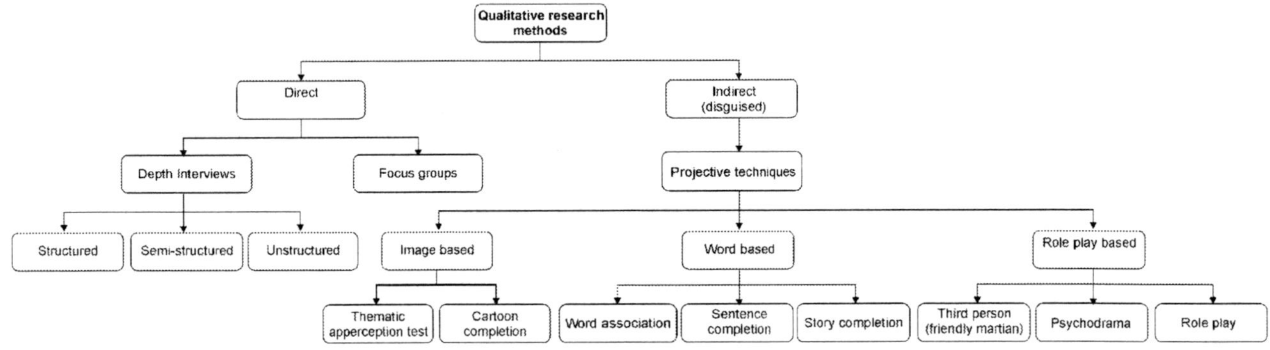

2 Qualitative data collection tools

We do not plan to consider in detail the entire set of methods shown in the diagram above but rather cover the main approaches for you to consider and direct you to others via the GMP VLE.

Assessment advice

The context of your own research will determine which method you employ and so we will present within this chapter the main methods, but you will need to supplement your reading depending on the methods you choose. The Anglia Ruskin digital library should be used extensively for the purposes of developing additional insight into the various data collection methods.

2.1 Interviews

Interviews are perhaps the most common of all qualitative methods.

> **Definition**
>
> According to Robson (2009) an '*interview is a conversation with a purpose. Interviews carried out for research or enquiring purposes are a very commonly used approach, possibly in part because the interview appears to be a quite straightforward and non-problematic way of finding things out.*'

In practice, interviews are likely to entail more than just a conversation. Interviews often involve a set of assumptions and understandings about the situation which are not normally associated with a casual conversation (Denscombe, 2003).

2.1.1 Planning interviews

The factors to consider when planning an interview are as follows.

(a) **Who should the respondent be?**

    (i)   The kind of person depends on the subject being discussed. It may be a consumer interview for discussion of consumer goods or an executive interview for discussing industrial buying.

    (ii)   The number of people undergoing depth interviews in the course of the research should be considered in the light of the time they take. 10-20 is usually more than enough (we will revisit this later in section 3).

    (iii)   Respondents for consumer interviews are pre-recruited and asked to agree to the interview.

(b) **What type of interview?** Although depth interviews are usually one-to-one, there may be more than one respondent and there may also be an informant, there to give information about tangible things (eg how big the organisation's purchase budget is) but not about his own attitudes.

(c) **How long should it be?** Genuine depth interviews interpret the meanings and implications of what is said and can therefore take some time. By contrast, a mini-depth interview may take only 15 minutes, because it can focus on one, predefined topic like a pack design.

(d) **How structured should it be?** It can be totally open-ended, ranging over whatever topics come up, or it can be semi-structured with an interview guide and perhaps the use of show material.

(c) **What material should be used?** The type of material that is commonly used includes mock-ups or prototypes, storyboards or concept boards, narrative tapes and other projective devices (we will return to projective techniques later).

(f) **Where should the interview take place?** Usually at home or in the workplace.

Global case study

Interviewing via internet enabled tools such as Skype is likely to become more popular particularly when conducting global research where there are cost contraints.

Bell (2005) draws on Grebenik and Moser (1962) in suggesting that the various types of interview can be described in terms of 'a continuum of formality'. Bell explains that '*at one extreme is the completely formalised interview where the interviewer behaves as much like a machine as possible. At the other is the completely informal interview in which the shape is determined by individual respondents. The more standardised the interview, the easier it is to aggregate and quantify the results.*'

We have prepared the following summary of the key types of interview, drawn from various sources.

Type of interview	Characteristics
Fully structured	<ul><li>Pre-determined or standard set of questions asked</li><li>The wording of the questions and their order are set</li><li>The respondent is invited to limited-option responses, ie pre-coded answers</li><li>Responses are recorded on a standardised schedule</li><li>The questions should be read out in the same tone of voice so as not to suggest any bias, ie there is what is called 'stimulus equivalence'</li><li>The data collected can be statistically interpreted but large samples are required to provide adequate confidence levels</li></ul>
Semi-structured	<ul><li>Questions are worked out in advance, but depending on the context of the 'conversation', the interviewer may adapt the schedule from one interviewee to the next as follows:<ul><li>– Change the wording of the questions</li><li>– Provide explanations</li><li>– Omit particular questions which seem inappropriate with particular interviewees</li><li>– Add further questions</li></ul></li><li>The responses are open-ended and there is more emphasis on the interviewee elaborating points of interest</li><li>Responses are recorded by taking notes, or where the interviewee agrees, by tape recording</li></ul>
Unstructured	<ul><li>There is no list of questions</li><li>The interviewer has a general area of interest and concern<ul><li>– allows the conversation to develop within this area, ie is non-directive</li><li>– should be as unintrusive as possible</li></ul></li><li>Also referred to as 'in depth' interviews. The aim is 'discovery', rather than 'checking'</li></ul>

(Robson, 2009; Denscombe, 2003; Saunders *et al* 2009)

Collis and Hussey (2003) suggest that structured interviews are associated with a positivistic methodology, whereas unstructured interviews are more in keeping with a phenomenological approach.

Where an unstructured format is used, within a phenomenological paradigm, the researcher will need adequate interpersonal skills to be able to build the confidence of the interviewee so as to be able to discuss and explore the relevant issues on an **in-depth** and **open** manner.

Easterby-Smith *et al* (2002) caution that '*although interviewing is often claimed to be 'the best' method of gathering information, its complexity can sometimes be underestimated. It is time consuming to undertake interviews properly, and they are sometimes used when other methods may be more appropriate.*'

Interviews are a very flexible research tool and can be used as the primary or only approach in a study, as in a survey or many grounded theory studies. They lend themselves well to use in combination with other methods, in a multi-method approach. In particular, a case study might even employ some kind of relatively formal interview to complement participant observation. An experiment could often usefully incorporate a post-intervention interview to help bring the participant's perspective into the findings (Robson, 2009).

Saunders *et al* (2009) relates the various types of interview to exploratory, descriptive and explanatory studies.

Type of study	Type of interview
Exploratory study	• In-depth interviews can be very helpful to 'find out what is happening [and] to seek new insights (Robson, 2009) • Semi-structured interviews also lend themselves to exploratory studies
Descriptive study	• Structured interviews can be used as a means to identify general patterns
Explanatory study	• Semi-structured interviews may be used in order to understand the relationship between variables • Structured interviews can be used in a statistical sense

(Saunders *et al*, 2009)

The various types of interview may also be used to compliment each other.

In-depth interviews may be used initially to identify variables. Data gathered is then used to design **questionnaires** or **interview schedule.**

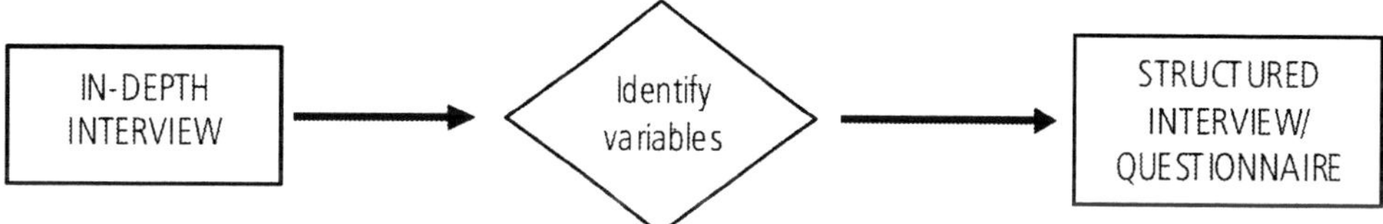

(Saunders *et al*, 2009)

Semi-structured interviews may be used to explore and explain themes that have emerged from the use of your questionnaire.

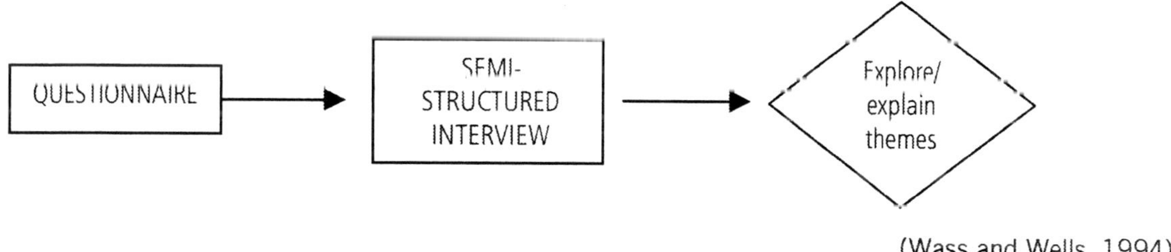

(Wass and Wells, 1994)

Semi structured and **in-depth interviews** used as a means of validating findings from questionnaires administered.

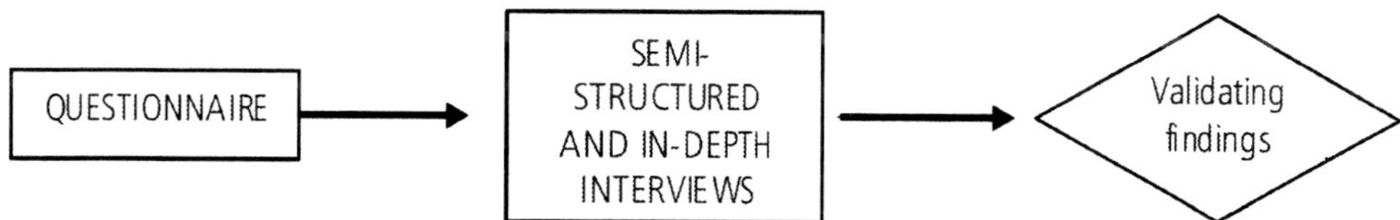

Saunders *et al* (2009) define interviewer bias as being '*where the comments, tone or non-verbal behaviour of the interviewer creates bias in the way that interviewees respond to the questions being asked*'.

For quantitative research, freedom from interviewer bias is essential. In a structured interview this includes ensuring that questions are communicated in a consistent manner from one interviewer to another. The perceptions and standpoint of the interviewer should not impact on the interviewee's response, nor the way it is recorded. Bell (2005) warns that '*the same question put by two people, but with different emphasis and in a different tone of voice can produce very different responses*'.

Assessment advice

This will not be a problem within your research because you are going to be working alone rather than within a team of researchers.

With qualitative research, such as the conduct of unstructured interviews, the approach is more often with a view to discovering the world of the interviewee. The interviewee has more 'space' to express his or her views and hence there should inherently be less scope for the interviewer to influence the interviewee's response (Easterby-Smith *et al*, 2002).

Where a team of interviewers is employed, there might be serious bias but this is likely to be revealed in the data analysis. But if (only) one researcher conducts a set of interviews, the bias may be consistent and therefore go unnoticed. In practice it is difficult to avoid this bias completely. Awareness of the problem plus constant self-control might be helpful (Bell, 2005).

Bell (2005) warns against the case one can lapse into in leading an interviewee. Collis and Hussey (2003) suggest that '*it is difficult to predict or measure potential bias, but you should be alert to the fact that it can distort your data and hence your findings*'.

2.1.5 Interviewer effect

The appearance as well as the behaviour of the interviewer is likely to impact on the response of the interviewee. These include:

- Gender, age and ethnic origins of the interviewer
- The personality of the interviewer
- The occupational status of the interviewer
- The appearance and manners of the interviewer
- The neutrality of the interviewer

(Denscombe, 2003)

Easterby-Smith *et al* (2002) refer to the work of Jones (1985) who points out that interviewees will 'suss out' what researchers are like, and make judgements from their first impressions about whether the interviewer can be trusted or whether they might be 'damaged' in some way by the information they provide. Such suspicions do not necessarily mean that interviewees will refuse to be interviewed, but it might mean, that they just try to get the interview over as quickly as possible.

It is common sense that you should dress in an appropriate way which is in keeping with the interview setting as well as the expectations of the interviewee. Thus, if you are interviewing a CEO of a public listed company, you are likely to wear smart business wear. If you are interviewing workers on a building site, you are likely to don a helmet, boots and informal clothes. You should however avoid being too blatant in mirroring the appearance of your interviewees or they may be insulted.

The issue of gender bias is highlighted by the work of Rosenthal (1966), as cited in Collis and Hussey (2003), which suggests different data collection results depending on whether the interviewer and interviewee are either male or female.

- Female subjects tend to be treated more attentively and considerately than male subjects.

- Female researchers tend to smile more often than male researchers.

- Male researchers tend to place themselves closer to male subjects than do female researchers.

- Male researchers tend to show higher levels of body activity than do female researchers; when the subject is male, both male and female researchers tend to show higher levels of body activity than they do when the subject is a female.

- Female subjects rate male researchers as more friendly and as having more pleasant and expressive voices than female researchers.

- Both male and female researchers behave more warmly towards female subjects than they do towards male subjects, with male researchers the warmer of the two.

Global case study

It is likely that the above guidelines will be culturally dependent and so could vary within your research setting.

2.1.6 Interview skills

We have looked at the risk of interviewer bias. One of the factors that might minimise the incidence of interviewer bias is the application of good interviewing skills by the researcher. Sound interviewing skills are also likely to contribute to your success as a manager in providing sound data for effective decision making.

Denscombe (2003) has explored the key elements of interviewing skills, which we have summarised here.

Skill	Key points
Attentiveness	• The interviewer needs to maintain the thread of the discussion and continue to listen closely while writing notes, reading body language or checking the tape recorder is working.
Sensitivity to feelings of interviewees	• Social courtesy is one aspect to be considered. • An aim is to get the best out of an interview. • Should foster a situation where the interviewee is more open with providing data
Able to tolerate silences	• This gives the interviewee the chance to develop their responses. • Experience should help the interviewer to avoid fearing that silence could lead to the failure of the interview.
Adept at using prompts	• These get the interviewee going without being led. • Nudges the interviewee gently into revealing their knowledge or thoughts on a specific point.
Adept at using probes	• Applied where the interviewer would like to delve more deeply into a topic rather than let the discussion flow on to the next point. • This may be necessary because the interviewee's response appears incomplete or inconsistent. • Should be subtle and avoid seeming aggressive.

Skill	Key points
Adept at using checks	• Involves summarising what has been said at strategic points during the interview. • Allows the interviewee to confirm or correct the details of the discussion.
Allows the interviewees to have their say	• Situations where a dominant personality influences the discussion should be avoided. • Encourages the interviewee to openly voice his or her response.
Non-judgmental approach	• The interviewer should avoid allowing his personal values to intrude or to show emotions like disgust, surprise or pleasure through facial gestures.

(Denscombe, 2003)

Easterby-Smith *et al* (2002) has developed a useful array of the various ways probes can be used.

1   The **basic probe** simply involves repeating the initial question and is useful when the interviewee seems to be wandering off the point.

2   **Explanatory probes** involve building onto incomplete or vague statements made by the respondent. Ask questions such as 'What did you mean by that?', 'What makes you say that?'

3   **Focused probes** are used to obtain specific information. Typically one would ask the respondent 'What sort of...?'

4   **The silent probe** is one of the most effective techniques to use when the respondent is either reluctant or very slow to answer the question posed. Simply pause and let them break the silence.

5   **Drawing out** is a technique that can be used when the interviewee has halted, or dried up. Repeat the last few words she said, and then look expectant or say, 'Tell me more about that', 'What happened then?'

6   **Giving ideas or suggestions** involves offering the interviewee an idea to think about – 'Have you thought about...?' 'Have you tried...?' 'Did you know that....?' 'Perhaps you should ask Y...'

7   **Mirroring or reflecting** involves expressing in your own words what the respondent has just said. This is very effective because it may force the respondent to rethink her answer and construct another reply which will amplify the previous answer – 'What you seem to be saying/feeling is...'

To avoid bias, probes should never lead. An example of a leading probe might be 'So you would say that you were really satisfied?' Instead the interviewer should say, 'Can you explain a little more?' or 'How do you mean?'

Robson (2009) provides a useful description of an interview process.

Step 1   **Introduction includes:**
- Personal introduction
- Explanation of purpose
- Give assurance on confidentiality
- Discuss procedures for making a record of the interview

Step 2   **Warm up**. Use of easy, non-threatening questions to settle interviewer and interviewee.

Step 3   **Interview**. Conduct main body of the interview.

Step 4   **Cool off**. Use straightforward questions to defuse any tensions that may have arisen.

Step 5   **Closure**. Say thank you and goodbye. Accompany interviewee to door with courtesy.

Denscombe (2003) suggests that as '*an information gathering tool, the interview lends itself to being used alongside other methods as a way of supplementing their data, adding detail and depth*'.

Use	Details
Preparation for a questionnaire	• Fine-tune the questions and concepts that will appear in a questionnaire. • Gain the detail and depth required to ensure that the questionnaire asks valid questions.
Follow-up to a questionnaire	• Pursue in greater detail and depth any interesting lines of enquiry thrown up by a questionnaire administered. • To complement the data gathered from a questionnaire.
Triangulation with other methods	• Can be used as a means of corroborating data gathered by other approaches.

(Denscombe, 2003)

LSE's Media and Communications department has a longitudinal research project titled EU Kids Online. In their information for parents section they have a list of resources including interview schedules and focus group discussion guides. They are worth referring to in order to see these real life data collection tools. The resources are available at:

www2.lse.ac.uk/media@lse/research/EUKidsOnline/EU%20Kids%20I%20(2006-9)/Interview%20schedules/Home.aspx

2.1.8 Advantages and disadvantages of interviews

The strengths of interviews include the following.

(a)   **Longitudinal information** (such as information on decision-making processes) can be gathered from one respondent at a time, thereby aiding clarity of interpretation and analysis.

(b)   Intimate and **personal material** can be more easily accessed and discussed.

(c)   Respondents are **less likely to confine themselves** simply to reiterating socially acceptable attitudes.

There are, however, **disadvantages** of interviews.

(a)   They are **time consuming** to conduct and to analyse. If each interview lasts between one and two hours, a maximum of three or four per day is often all that is possible. There is also a knock on effect with regards to the time taken to analyse findings. Every hour of an in depth interview can lead to around 30 pages of typed A4 paper!

(b)   They are more **costly** than group discussions (due to time and travel expenses).

(c)   There is a temptation to begin treating depth interviews as if they were simply another form of questionnaire survey, thinking in terms of quantitative questions like 'how many' rather than qualitative issues like 'how', 'why' or 'what'.

Activity 1

Why do you think researchers tend to record interviews?

Which do you think is preferable: audio or video tape?

2.2 Focus groups

Focus groups are widely used in the fields of market research and advertising and this popularity has spread to social research more generally in recent years (Denscombe, 2003).

According to Collis and Hussey (2003) '*focus groups are normally associated with a phenomenological methodology. They are used to gather data relating to the feelings and opinions of a group of people who are involved in a common situation*'.

Focus groups draw on the effects of group dynamics to provide useful research data. '*The explicit use of the group interaction to produce data and insights that would be less accessible without the interaction found in a group.*' (Morgan, 1997).

Collis and Hussey (2003) provide a process for forming a focus group which we have summarised in the following diagram.

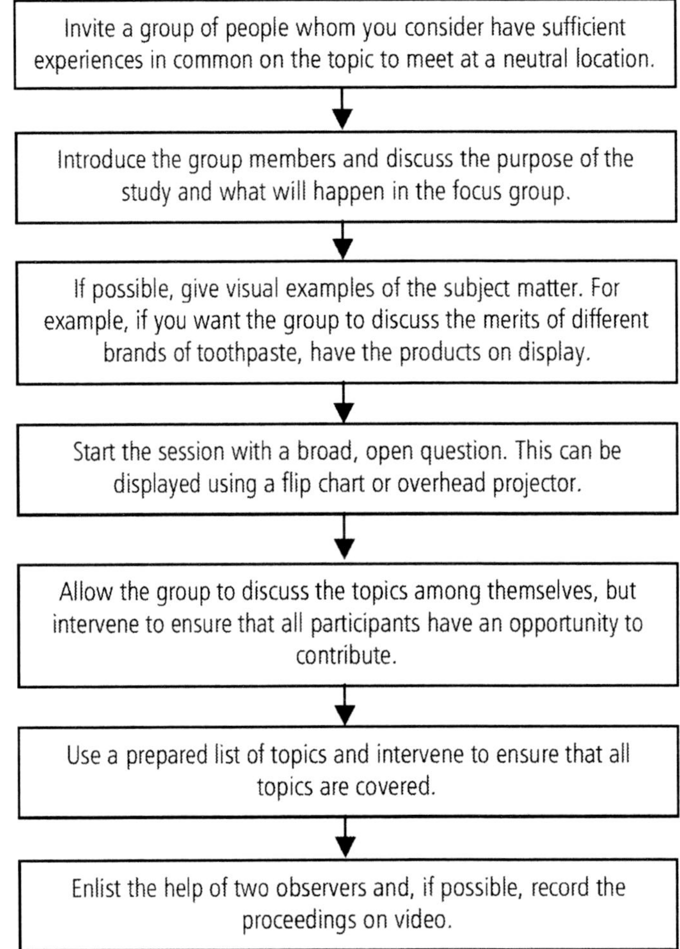

(Adapted from Collis and Hussey, 2003)

Various techniques are used to encourage and record responses such as the use of cards by participants to jot down their thoughts or ideas. These can then be collected by the facilitator for further discussion. Sometimes these are displayed on flipcharts using a spray on adhesive.

According to Collis and Hussey (2003), '*focus groups combine both **interviewing** and **observation**. They are often used in pilot studies to develop a **questionnaire** or **interview schedule** for quantitative study.*' However, as Collis and Hussey explain, '*the data generated from a focus group is qualitative*'.

Focus groups concentrate on discussion of chosen topics in an attempt to find out attitudes. They do have limitations despite advantages such as the ability to observe a whole range of responses at the same time.

These are useful in providing the researcher with qualitative data.

Focus groups usually consist of 8 to 10 respondents and an interviewer taking the role of group moderator. The group moderator introduces topics for discussion and intervenes as necessary to encourage respondents or to direct discussions if they threaten to wander too far off the point. The moderator will also need to control any powerful personalities and prevent them from dominating the group.

The researcher must be careful not to generalise too much from such small scale qualitative research. Group discussion is very dependent on the skill of the group moderator. It is inexpensive to conduct, it can be done quickly and it can provide useful, timely, qualitative data.

Focus groups are often used at the early stage of research to get a feel for the subject matter under discussion and to create possibilities for more structured research. Four to eight groups may be assembled and each group interviewed for one, two or three hours.

When planning qualitative research using focus groups, a number of factors need to be considered.

(a)   **Type of group** A standard group is of 8 to 10 respondents, but other types may also be used.

(b)   **Membership.** Who takes part in the discussion depends on who the researcher wants to talk to (users or non-users, for instance) and whether they all need to be similar (homogenous).

(c)   **Number of groups.** Having more than twelve groups in a research project would be very unusual, mainly because nothing new would come out of a thirteenth one!

(d)   **Recruitment.** Usually on the basis of a quota sample: respondents are screened by a short questionnaire to see whether they are suitable. In order to persuade them to join in, the members are usually given an incentive plus expenses.

(e)   **Discussion topics.** These will be decided by the researcher with regard to the purpose of the group discussion, that is the data that is required. There should be a number of topics since the interviewer needs to be able to restart discussion once a topic has been fully discussed.

2.2.1 Discussion guide

A discussion guide (also known as a topic guide) is the guide prepared by a depth interviewer or focus group moderator to guide the topics under discussion. Topic guides can take many different forms, according to clients' preferences and the needs of the research, from looser lists of subject areas to be covered to more strictly structured lists of specific question areas.

The general format is to have three phases.

(a)   **Introduction**, welcoming the group, explaining the purpose of the meeting and outlining the topics that will be discussed, and introducing the participants or getting them to do so themselves.

(b)   **Discussion**, which may have several themed sub-phases and involve product trial.

(c)   **Summary** and thanks.

Global case study

SAMPLE DISCUSSION GUIDE: HIGH FIBRE MICROWAVE PIZZA

Introduction

Introduce self, note ground rules and mention taping.

Warm up: Go around the table and state name and what types of pizzas you buy (briefly).

Discussion

Discuss pizza category: What's out there? What is most popular? What's changed in your pizza eating habits in the last five years?

When cooking pizza at home, what kinds do you make? (frozen, chilled, baked, microwaved, etc.) Any related products?

Probe issues: Convenience, costs, variations, family likes and dislikes

Probe issues: Any nutritional concerns?

Present concept: Better nutritional content from whole wheat and bran crust, high in dietary fibre. Strong convenience position due to microwavability. Competitive price. Several flavours available. Get reactions.

Probe: Is fibre a concern? Target opportunity for some consumers?

Probe: Is microwave preparation appealing? Concerns about browning/sogginess/crispness?

Taste and discuss representative prototypes. Discuss pros and cons. Probe important sensory attributes. Reasons for likes and/or dislikes.

Review concepts and issues. Ask for clarification.

Ask for new product suggestions or variations on the theme.

Last chance for suggestions. False close (go behind mirror).

If further discussion or probes from clients, pick up thread and restart discussion.

Summary

Close, thanks, distribute incentives, dismissal.

2.2.2 Moderator

Key to the success of a focus group will be the skill of the person running it, usually called the **moderator**.

(a)   The moderator needs to be able to **build a rapport** between a group of people who have never met before.

(b)   He or she needs to make sure that **everyone gets an opportunity** to speak.

(c)   The moderator needs to ensure that the discussion **stays focused** on the topics at hand.

(d)   The moderator must be **sensitive to the mood** of the group throughout the discussion. It is likely to change a number of times even during a relatively short meeting.

Dillon *et al* (1994) believes that the moderator plays the same role as a therapist does in group therapy. As such they should discuss topics within a time period but not in any set order. The role of a moderator is not to lead or direct the discussion but to keep it focused on the topic of interest.

Moderator discussion guides should be flexible enough to be altered as the discussion progresses.

Self-development

Next time you are in a meeting see if you can sense the changes in mood. You are most likely to notice a change when the meeting has gone on a bit too long and people start to want to leave! Look for tell-tale signs like body language and tone of voice, as well as changes in the quality of discussion.

Wilson (2006) draws attention to the typical thinking process of a focus group.

(a)   Initially there will be **anxieties and doubts**: participants won't know exactly why they are there and what is expected of them. The moderator needs to be aware of this and set their minds at rest: perhaps even ask them to share their doubts.

(b)   Participants will **wish to feel included** so it is important to get contributions from each member early on.

Research Methods for Global Marketing Practice

(c)   Especially because they are strangers, the group members are likely to want to **establish their own status**: what their experience of the matter under discussion is, why they should be listened to. This tends to die down after a while and participants are more interested in sharing and **relating to each other**. Depending on the individuals, however, intervention from the moderator may be required to prevent one or two people dominating the discussion.

(d)   People are keen to feel that their **opinions are valued**: the moderator needs good listening skills (good eye contact, asking for points made to be developed).

(e)   Sooner or later people will start to **wish to leave**, so the moderator needs to give indications of how much ground has been covered and what is left to be covered at regular intervals.

Advantages and disadvantages of focus groups

The **key advantages** of focus groups include the following.

(a)   The group environment with 'everybody in the same boat' can be **less intimidating** than other techniques of research which rely on one-to-one contact (such as depth interviews).

(b)   What respondents say in a group often **sparks off experiences** or ideas on the part of others.

(c)   **Differences between consumers** are highlighted, making it possible to understand a range of attitudes in a short space of time.

(d)   It is **easier to observe groups** and there is more to observe simply because of the intricate behaviour patterns within a collection of people.

(e)   **Social and cultural influences** are highlighted.

(f)   Groups provide a **social context** that is a 'hot-house' reflection of the real world.

(g)   Groups are **cheaper and faster** than depth interviews.

(h)   **Technology** may help to facilitate and add value to the process.

    (i)   Group discussions can be **video tape recorded** for later analysis and interpretation, or they may even be **shown 'live'** to the client via CCTV or webcam.

    (ii)   In business-to-business situations it may be possible to use **video-conferencing**, enabling opinions to be sought from a wider variety of locations.

    (iii)   **Forums** and **chat rooms** on the web can be used.

The principal **disadvantages** of groups are as follows.

(a)   Group processes may **inhibit some people from making a full contribution** and may encourage others to become exhibitionistic.

(b)   Group processes **may stall** to the point where they cannot be retrieved by the moderator.

(c)   Some groups may **take a life of their own**, so that what is said has validity only in the short-lived context of the group.

(d)   It is not usually possible to identify **which group members said what**, unless the proceedings have been video recorded.

2.3 Projective techniques

Projective techniques attempt to draw out attitudes, opinions and motives by a variety of methods.

Many interview techniques rely on the assumption that you need only to ask people and they will tell you what you want to know. This is not always the case. People may respond differently to how they would act. People may tell you what they think you want to hear or give a different answer because their true answer may reflect badly on them or because they consider it too personal.

Alternatively, people may find difficulty in articulating their motives which lie buried deep within the sub-conscious mind. To overcome problems associated with articulating complex or sub-conscious motives, researchers have borrowed techniques developed by psychologists in their studies of people with mental health problems who may have difficulty explaining why they do things.

These techniques are referred to as **projective techniques**. Attitudes, opinions and motives are drawn out from the individual in response to given stimuli.

A number of techniques might be employed.

(a)   **Third person,** or 'friendly Martian' as it is sometimes called, is designed to get the respondent talking about issues which do not interest them. The researcher asks the respondent to describe what someone else might do (a friendly Martian). For example, if someone wanted to buy a house, what do they need to do? Can you describe the steps they would need to take?

(b)   **Word association** is based on an assumption that if a question is answered quickly, it is spontaneous and sub-conscious thoughts are therefore revealed. The person's conscious mind does not have time to think up an alternative response.

(c)   **Sentence completion** is a useful way to get people to respond quickly so that underlying attitudes and opinions are revealed.

- Men who watch football are ....?
- Women wear red to ....?
- People who Morris dance are ....?

(d)   In **thematic apperception tests** (TAT tests), people are shown a picture and asked to describe what is happening in the picture. They may be asked what happened just before or just after the picture. It is hoped that the descriptions reveal information about deeply held attitudes, beliefs, motives and opinions stored in the sub-conscious mind.

(e)   **Story completion** allows the respondent to say what they think happens next and why.

(f)   **Cartoon completion** is often used in competitions. There are usually speech balloons which need to be completed. A comment may be present in one and another left blank for the respondent to fill in.

(g)   **Psychodrama** consists of fantasy situations. Respondents are often asked to imagine themselves as a product and describe their feelings about being used. Sometimes respondents are asked to imagine themselves as a brand and to describe particular attributes.

2.3.1 Problems and the value of projective research methods

There are a few problems associated with projective techniques.

(a)   Hard evidence of their validity is lacking. Highly exotic motives can be imputed to quite ordinary buying decisions. (One study concluded that women preferred spray to pellets when it came to killing cockroaches because being able to spray the cockroaches directly and watch them die was an expression of hostility towards, and control over, men!)

(b)   As with other forms of intensive qualitative research, the samples of the population can only be very small, and it may not be possible to generalise findings to the market as a whole.

(c)   Analysis of projective test findings – as with depth interviews – is highly **subjective** and prone to bias. Different analysts can produce different explanations for a single set of test results.

(d)   Many of the tests were not developed for the study of marketing or consumer behaviour, and **may not therefore be considered scientifically valid** as methods of enquiry in those areas.

(e)   There are **ethical problems** with 'invasion' of an individual's subconscious mind in conditions where he is often not made aware that he is exposing himself to such probing. (On the other hand, one of the flaws in projective testing is that subjects may be all too well aware of the nature of the test. The identification of sexual images in inkblots has become a standard joke.)

The major drawback with projective techniques is that answers given by respondents require considerable and **skilled analysis and interpretation**. The techniques are most valuable in providing **insights** rather than **answers** to specific research questions.

However, motivational research is still in use. Emotion and subconscious motivation is still believed to be vitally important in consumer choice, and qualitative techniques can give marketers a deeper insight into those areas than conventional, quantitative marketing research.

Since motivational research often **reveals hidden motives** for product/brand purchase and usage, its main value lies in the following.

(a)   Developing **new promotional messages** which will appeal to deep, often unrecognised, needs and associations

(b)   Allowing the **testing of brand names**, symbols and advertising copy, to check for positive, negative and/or irrelevant associations and interpretations

(c)   Providing **hypotheses which can be tested** on larger, more representative samples of the population, using more structured quantitative techniques (questionnaires, surveys).

There is debate within the marketing literature about whether projective techniques are a qualitative or quantitative research tool as they are used in both research designs. For instance they can be used successfully within quantitative questionnaires (Bradley, 2007, Dillon *et al*, 1994). Wilson (2006) on the other hand directly states that projective techniques are used in group discussions and depth interviews.

2.4 Online qualitative research

In less than a decade a third of US research agencies' revenues has shifted to online research. Rather than the research industry being threatened by the new digital environment, Cooke and Buckley (2008) believe that it offers an opportunity to develop new research approaches. The researchers from agency GFK NOP identified Web 2.0 as one example of where innovative methods can be used to explore changing social environments. They clearly define Web 2.0 as follows:

'Web 2.0 refers to the new generation of tools and services that allow private individuals to publish and collaborate in ways previously available only to corporations with serious budgets, or to dedicated enthusiasts and semi-professional web builders.'

It is built around the concept of social software that enables people to collaborate and form online communities. Online communities combine one to one (email, instant messages), one to many (web pages, blogs) and many to many (social networking sites eg Wikis, Facebook, Second Life) communication modes.

The growth in social networking is significant for market researchers because within society a population is growing of people who are more willing to record and share their experiences with friends and other community members for evaluation. This evaluation forms the basis of individuals' reputations and possibly self concept. Social networks also highlight human tribal behaviour. Earls (2003) suggested that market researchers have in the past overlooked the most important part of what it means to be human – we are herd animals. He argued that market researchers should study the interaction between individuals to make informed decisions about consumer behaviour (Earls, 2003).

Social software which supports group interaction can include, blogs, wikis, podcasts, P2P file sharing, virtual worlds and social networks. Additional content is often generated following a 'mash up' where information is mixed from a number of disparate sources to produce creative data. The key issue for researchers to grasp according to Cooke and Buckley (2008) is that social networks are formed voluntarily and sub groups develop as experiences are shared and reputations are built. Within these online environments, members set the agendas and conversations and therefore selectively only enter groups where there is a common interest.

For the market researcher, social networks can be created or existing networks used where community members are already interested in the subjects being investigated. Both qualitative and quantitative research is possible and provides a wealth of opportunity for not only increasing response rates as a result

of precision sample selection and respondent interest, but the ability to hold more sophisticated extended focus groups and interactive panels. (Although to describe social network research in these two traditional terms could be considered to undermine their true value and future opportunity.)

Activity 2

Imagine you are investigating a new form of refrigerator which is highly environmentally friendly with a negligible carbon footprint. You know through secondary research that this would appeal to a growing group of ecologically aware consumers but because the production costs are high, you need to conduct some detailed research into the price that target consumers would be willing to spend.

How might social networks help with this research?

Global case study

The virtual world Second Life has a growing number of established 'real-world' brands using its communities to pilot test product launches, idea generation, concept testing and customer experience research. Second Life enables users to create alternative realities from scratch and so they could technically be living a double life, one within the real world with a virtual existence alongside. As lifestyles are built within these virtual lives, brands are entering the virtual marketplace for members to purchase and conspicuously use in order to develop their virtual self concept and reputation. Dell for example enables Second Life users to build their own bespoke computer to use in their virtual life and then to even buy the finished product in the real world if they wish.

Lego similarly through their own website enables member users to build their own models from over 500 pieces, this equates to a massive design team to assist with NPD (New Product Development).

(Cooke, N. & Buckley, N., 2008)

3 Sampling issues in qualitative research

Sampling for qualitative research should be treated differently to sampling for quantitative research. The same overall systematic approach to sample selection should be used but with some very important differences.

In Chapter 3 we looked at the overall process of sampling. How it might need to be employed in qualitative research can be seen in the table below:

Sampling issues	Relevance to qualitative methods
1 **Who** do we need to research? (population of interest).	This is still relevant.
2 **Where** do we find them? (sampling frame).	This is still relevant.
3 How should we **select** individual respondents? (sampling method)	This is still relevant. Snowball sampling is very frequently used within qualitative research.

4	**How many** respondents do we need? (sample size).	Depth interviews may have fewer than ten respondents although 20 seems to be a universally accepted rule of thumb. (Dillon, 1994).
5	Will research respondent views be **representative** of the views of everybody else in that situation?	We should not think of generalisability when conducting qualitative research.

4 Analysing qualitative data

Palton (1990) stated that:

'The purpose of qualitative inquiry is to produce findings. The process of data collection is not an end in itself. The culminating activities of quantitative inquiry are analysis, interpretation, and presentation of findings. The challenge is to make sense of massive amounts of data, reduce the volume of information, identify significant patterns, and constructing a framework for communicating the essence of what the data reveal.'

4.1 What is qualitative data?

Quantitative data has the following features.

- It is based on **numbers** (categorical or quantifiable).
- There are a limited number of **standard** responses.
- Analysis of quantitative data involves the use of **graphical** and **statistical** techniques.

Qualitative data, on the other hand, has the following features.

- It is based on **words.**

- There are an **unlimited number** of **non-standard** responses.

- Analysis of qualitative data involves a technique known as **conceptualisation** which we will be investigating in this section of the chapter. It is also known as **content analysis**.

4.2 Variety of strategies and approaches

Qualitative data is complex in nature and therefore it is difficult to provide a set of rules to be followed in its analysis if we wish to analyse and draw conclusions from our research. There are therefore many ways in which this type of data can be analysed.

This section will provide a brief overview of qualitative data analysis and the GMP VLE will direct you to other academic texts which you should refer to if you would like to use those techniques to analyse your data. The key consideration is to apply the most useful and appropriate techniques.

Remember that there are no numbers (unless you **quantify** your data), and the nature of qualitative data is such that you will need to **plan** your work very carefully before commencing. You will need to approach the analysis in a very **methodical manner**. You need to bear in mind your research objectives and assess the nature and usefulness of the data you have collected.

Saunders *et al* (2009), explain that the objective for adopting a systematic approach to analysing your data rigorously is to be able to draw verifiable conclusions.

Indeed Saunders *et al* (2009) advise that the data analysis process should really be considered up front at the time you are preparing your research proposal or terms of reference.

Palton (1990) argues that the lack of available rules does not mean that there are no guidelines to assist in analysing data. He suggests that:

- Applying guidelines requires judgement and creativity.
- Each quantitative study is unique and hence the analytical approach used will be unique.
- It depends on the skills, training, insights and capabilities of the researcher.
- The human factor is the great strength but also the great weakness of quantitative analysis..

4.3 Main approaches to qualitative data analysis

In general, there are three main approaches to analysing qualitative data.

- Quantification
- Deductive
- Inductive

Quantification

Categorical data cannot be measured numerically, but can be placed into categories (according to the specific characteristics that it possesses). Categorical data is also a type of **qualitative data** which can be **counted** to establish the frequency of occurrence of certain categories. We will also return to this in Chapter 5.

Quantification of qualitative data can complement the principal methods of analysis that you are using when conducting your research. For example, you could include graphical representations of your data in a supplement or an appendix at the end of your research report.

Global case study

The table below shows the results of a series of semi-structured interviews with twelve managers in an organisation. The objective of this research was to evaluate the extent to which this organisation had adopted the human resources initiatives shown below.

In order to **quantify** the results of the interviews, the number of times each of the HR initiatives were referred to were counted and then tabulated as shown.

HR initiatives	HRM practices identified by senior/line managers (n = 12)													
	1	2	3	4	5	6	7	8	9	10	11	12	Total	%
Developing flexible staff base														
Restructuring			✓										1	8
New style (professional) contracts	✓										✓		2	17
Develop a skills mix														
Effective staff development														
Creation of Professional Development Unit	✓				✓								2	17
Appointment of Professional Development Tutor														
Funding staff development at 2% of payroll	✓				✓								2	17
Performance appraisal for all	✓			✓									2	17
Goal-orientated rewarded and appraisal														
Performance-linked pay scheme			✓		✓						✓		3	25
Generating staff commitment														
Ensuring equal opportunity			✓			✓							2	17
High-quality support staff structure														
Introduce quality improvement programme														
Customer care for personnel department														

HR initiatives	HRM practices identified by senior/line managers (n = 12)													
	1	2	3	4	5	6	7	8	9	10	11	12	Total	%
Harmonisation of staff terms and conditions														
Identification of HR needs from strategic plan														
Targeted appointments														
Development of robust HR MIS System			✓										1	8
Staff development review scheme	✓	✓	✓	✓	✓	✓	✓			✓	✓		9	75
Devolving staff development budgets									✓		✓		2	17
Investors in People		✓						✓					2	17
Induction programme			✓										1	8
Line management training			✓										1	8
In-house training programmes			✓										1	8
Nurturing staff from within			✓										1	8
Devolvement to line managers										✓	✓		2	17
Personnel/Policy Handbook			✓	✓									2	17
Formal structures to support HR function			✓										1	8
Transfer of payroll to personnel														
Harassment policy				✓									1	8
Grading review and promotion policy									✓		✓		2	17
Direct communication										✓			1	8
Communication audit			✓										1	8
Common Interest Groups			✓										1	8

Quantifying qualitative data (Source: Millmore, 1995)

Alternatively the following form of tabulation can be applied. A table is created with columns for the different kinds of respondents and rows for the research objectives. Here is a very basic example.

	Men	**Women**
Attitude towards watching sports		
Attitude towards playing sports		

The comments and quotes from the transcripts are then entered into the appropriate box. This makes it much easier to make comparisons and can give rise to some quantitative data ('six out of ten women said …').

Provided researchers are not allowed to add their own categories the task can be shared between several people with consistent results. On the other hand the method could be considered to be too inflexible: collected data that does not 'fit' anywhere might be discarded even though it is valuable.

4.4 Framework for analysing qualitative data

There is no standardised approach for analysing qualitative data, therefore there is a need to establish a **general framework** for dealing with this complex method of analysis. A typical framework might comprise various stages.

1   **Data categorisation**
2   **Data unitisation**
3   **Relationship identification**
4   **Theory/proposition development**

We shall look at each of these stages in turn.

4.4.1 Data categorisation

The first stage of the framework involves identifying a set of categories that reflect the data that you have collected. Because qualitative data is associated with a phenomenological approach and explains the deeper meanings of your research participants, the data collected is usually complex and comprehensive. To facilitate your interpretation of the data collected, it is therefore important to impose a workable structure to it.

The starting point is to annotate transcripts. As you might guess this entails the researcher categorising the items in a transcript by adding marginal comments, or perhaps using different coloured highlighter pens (or the equivalent in a word processor or spreadsheet). This leaves the actual data intact, so it is still possible to see how the full conversation flowed.

The categories that you select might be very different from those selected by another researcher since they will be determined by your research objective. For this reason, you must have a clearly defined set of objectives at the beginning of your project to inform and guide the analysis of your data.

The first stage of the framework will be completed when you have identified the set of categories relevant to your research and produced a description for each one.

4.4.2 Data unitisation

The second stage of the framework involves breaking down the data collected into 'bite-sized' **units** in order that they can be analysed further. A **unit** of data can be any of the following.

- Word(s)
- Sentence(s)
- Paragraph(s)

When units of data have been identified, you will need to attach 'labels' to them in order to show which category they are being allocated to. This can be done **manually** or by using suitable **qualitative data analysis software**. Often the labels are shortened into codes and a working code sheet is prepared. For example within some research on satisfaction with an airline, a large proportion of respondents cited the need for a good quality meal, flight times being kept and polite and attentive cabin crew as key areas leading to their satisfaction. The following codes might be adopted to represent these:

- FQ (good quality food)
- FT (flight on time)
- CP (crew politeness)

4.4.3 Manual data unitisation

On reviewing your qualitative data (or transcripts) you could manually mark up the margins with the relevant labels. This will enable you to identify to which **category** the different data units belong. When you have labelled all of your data, you can group units of data that have the same labels in order to produce a mass of information relating to individual categories. There are two ways in which this can be done.

- Data can be grouped together by copying units of data from transcripts which have the same labels and sticking them onto **data cards.** Each data card will therefore contain a mass of data units relating to a specific category. (Saunders *et al*, 2009)

- Alternatively, **index cards** can be used. A separate index card is needed for each category and each card gives a list of **references** which guide the researcher to all of the transcripts that contain information relating to the category in question. (Easterby-Smith *et al*, 2002)

As with categorisation, it is the objectives of **your** research that will determine the ways in which you break down your data into 'bite-sized' chunks (data units) and subsequently re-group them using data cards or index cards.

Spider diagrams or **mind maps**. The research issue is placed at the centre of a sheet of paper and the key themes that emerge and relevant quotes and comments from the transcripts are placed around it.

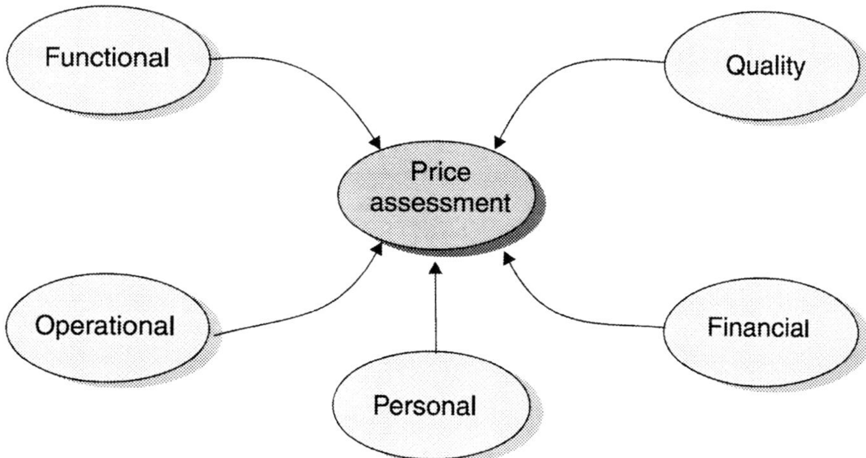

This makes it easier to include comments that don't 'fit' when using the tabulation method. The complexity of the interrelationships between items can be shown more clearly, via the placement of the comments and the use of interconnecting lines. Because the method is less rigid, there is no guarantee that two researchers would analyse the data in the same way, however.

We will consider the use of analysis software in order to unitise data at the end of this chapter.

4.4.4 Relationship identification

One of the main aims of qualitative data analysis is to identify any patterns or relationships that may be present within the data that you have collected. Within a deductive approach you are likely to have a set of research hypotheses and gathered data in order to test your hypotheses. Your data will be used to confirm or deny the relationship between the variables set out in your hypothesis.

Saunders *et al* (2009) and Miles and Huberman (2003) warn of the importance in hypothesis testing of considering alternative explanations. These may or may not be consistent with the pattern or relationship being tested. Saunders *et al* (2009) stress that your conclusions must be secure enough to withstand alternative explanations, bearing in mind that alternative explanations might exist.

Within an inductive approach, theory follows data collection. Hence you would examine the data you have collected to try to discern whether they answer your research question(s).

As your research progresses and as relationships emerge, you might consider it necessary to re-evaluate the set of categories that you developed originally and modify any that appear to be inappropriate. It is important to keep an up-to-date list of categories so that data units are labelled with the correct categories (Miles and Huberman, 2003).

Miles and Huberman (2003) suggest that a vital part of the reflections undertaken by the qualitative researcher will attempt to identify *'patterns and processes, commodities and differences'*. Denscombe (2003) suggests that when revisiting field notes, transcripts or text, you should be on the lookout for themes or interconnections that recur between the units and the categories that are emerging.

Denscombe (2003) also advises that as various explanations and themes emerge from the early consideration of data, you should go back to the field with these explanations and themes to check their validity against reality.

4.5 Tools supporting a deductive approach

The two key analytical tools used within a deductive approach are as follows.

- Pattern-matching
- Explanation-building

4.5.1 Pattern-matching

Pattern-matching is an analytical procedure that uses a theoretical framework in order to predict a pattern of **possible** outcomes. This procedure commences with using existing theory in order to develop a theoretical framework and then to establish whether this framework provides a convincing explanation of the results that you obtain while conducting your research. If the pattern of possible outcomes that were predicted (using the theoretical framework) match the research results that you obtained, you will have found an explanation for your results.

Yin (2009) advises that if your findings reveal one or more outcomes that have not been predicted by your explanation, you will need to seek an alternative one.

Medical research predicts that if people cut out cigarette smoking, average life expectation will increase. However, if findings show that average life expectation has not increased, alternative explanations may have to be found, eg people are probably still living unhealthy lifestyles or perhaps suffering from too much stress.

Here is a diagram summarising the above process.

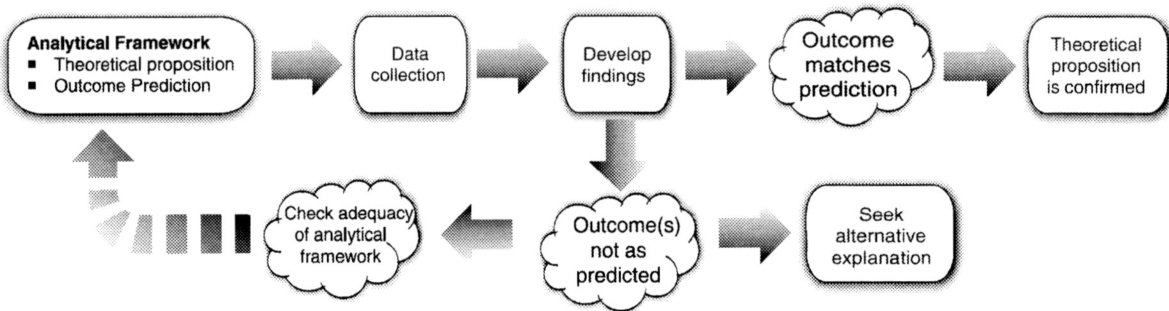

4.5.2 Explanation-building

This procedure involves trying to build up an explanation during data collection and analysis rather than putting a predicted pattern of outcomes to the test (as in pattern-matching). The main differences between pattern-matching and **explanation-building** are as follows.

- **Pattern-matching** develops a theoretical framework based on existing theory and seeks to test it.

- **Explanation-building** attempts to develop a theoretical framework using your findings during data collection and analysis. (Note that this procedure commences with a theoretical framework before data is collected and that this framework is fine-tuned by following the stages shown below).

Research Methods for Global Marketing Practice

1. Devise a theoretically-based proposition (based on relevant existing theory).

2. Compare the results of your data collection with your proposition.

3. Make adjustments to the original proposition (where necessary) in order to devise a revised proposition.

4. Collect more data.

5. Make further adjustments to the revised proposition (where necessary).

6. Repeat this process until you have built up a satisfactory explanation for the results that you have obtained.

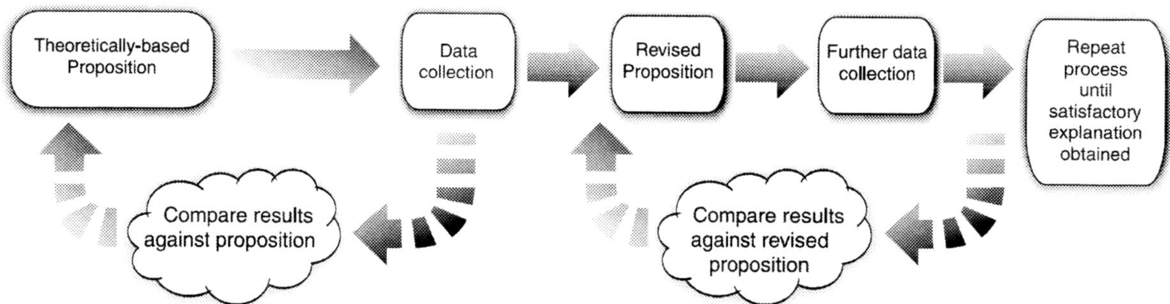

Activity 3

From the discussion above, do you think this highlights anything specific about the practical implications of using qualitative research?

4.6 Inductive approach

4.6.1 Objectives

One of the main differences between the deductive and inductive approaches to qualitative data analysis is the point at which the main categories are specified.

- With a **deductive approach**, this can be (although it does not have to be) before data collection begins (since it is based on existing theory)

- With an **inductive approach** the main categories emerge as data is collected and analysed.

4.6.2 Grounded theory

One of the most important qualitative research methods that adopt an inductive approach is grounded theory. We touched upon grounded theory in Chapter 1.

Grounded theory is most accurately described as a research method in which the theory is developed from the data (ie the theory is 'grounded' in the data) rather than the other way round. The use of grounded theory in research requires a high level of knowledge of your chosen research topic because you will not be making make use of any existing theories. You will also need to define your research objectives clearly before you start collecting your data.

In the grounded theory approach of Strauss and Corbin (1997), the following procedures are identified. Each of these procedures can be related to our framework for analysing qualitative data.

- **Open coding** – linked to 'categorising data' and 'unitising data'
- **Axial coding** – linked to 'identifying relationships'
- **Selective coding** – linked to 'developing theories/propositions'

4.6.3 Open coding

Open coding corresponds to the first and second stages in our framework for analysing qualitative data and involves breaking down the data you have collected into 'bite-sized' units and identifying any significant categories or themes that emerge from them. Each unit of data is given a label, with similar units being given the same labels. A unit of data might comprise a few words, a sentence or a number of sentences, or a paragraph (Saunders *et al*, 2009).

Grounded theory avoids making use of existing theory with categories and themes emerging from the data collected during your research.

It is worth bearing in mind that Saunders *et al* (2009) emphasise that grounded theory involves deriving meaning from the subjects being studied. Strauss and Corbin (1997) caution against code labels drawn too closely from existing theory or literature; they suggest use of the following sources:

- Terms emerging from your data collected.
- Actual terms used by participants, ie *in vivo* codes.
- Terms used in existing theory and literature.

4.6.4 Axial coding

Axial coding corresponds to the third stage in our framework for analysing qualitative data and involves identifying any relationships that may have emerged from the open coding process. Once the open coding process has been completed, data is restructured into various patterns in order to try to reveal potential relationships.

Principal categories and sub-categories can also be identified at the axial coding stage and you will probably be able to begin constructing propositions that might be used to aid the development of an explanatory theory in the selective coding process.

4.6.5 Selective coding

Selective coding corresponds to the final stage of the framework for analysing qualitative data and involves developing a grounded theory.

Once your data collection process has been completed, a number of principal categories and sub-categories will have emerged (in the axial coding process). Selective coding involves identifying the relationships which exist between these principal categories and identifying one of these principal categories as the central or core category.

If you are considering using the grounded theory approach in your qualitative research you should bear in mind the following points:

- It is very time-consuming.
- It is very intensive.
- It requires a high level of knowledge of the chosen research topic.
- The final results of your analysis may not necessarily reveal any significant results.

4.6.6 Using the deductive and inductive approaches together

You might begin a research project by adopting a deductive approach, that is to say by using existing theory in order to develop a theoretical framework on which to base your research. As your research progresses, it might become apparent that the framework you have adopted is not sufficiently relevant to

Research Methods for Global Marketing Practice

your research objectives. If this happens, you can re-analyse your data using an inductive approach in the hope that new, more relevant themes or categories will emerge. By combining the two different approaches to your data analysis, you might find that you are more successful in reaching your research objectives.

You may be conducting research into supermarket customer behaviour. You may have postulated that because a certain chain emphasises 'value for money' as its principle selling proposition that the majority of customers identify this as their main motivation for preferring to shop at this chain.

However, your analysis and findings reveal that price is a low consideration but instead people were attracted to the chain because it played piped background music.

You might therefore conduct further inductive research to try to discover why shoppers liked the music. This might lead you to exploring their lifestyles and perhaps more profound issues that the customers sought from their shopping experience and what meanings visits to the shops held for the people involved.

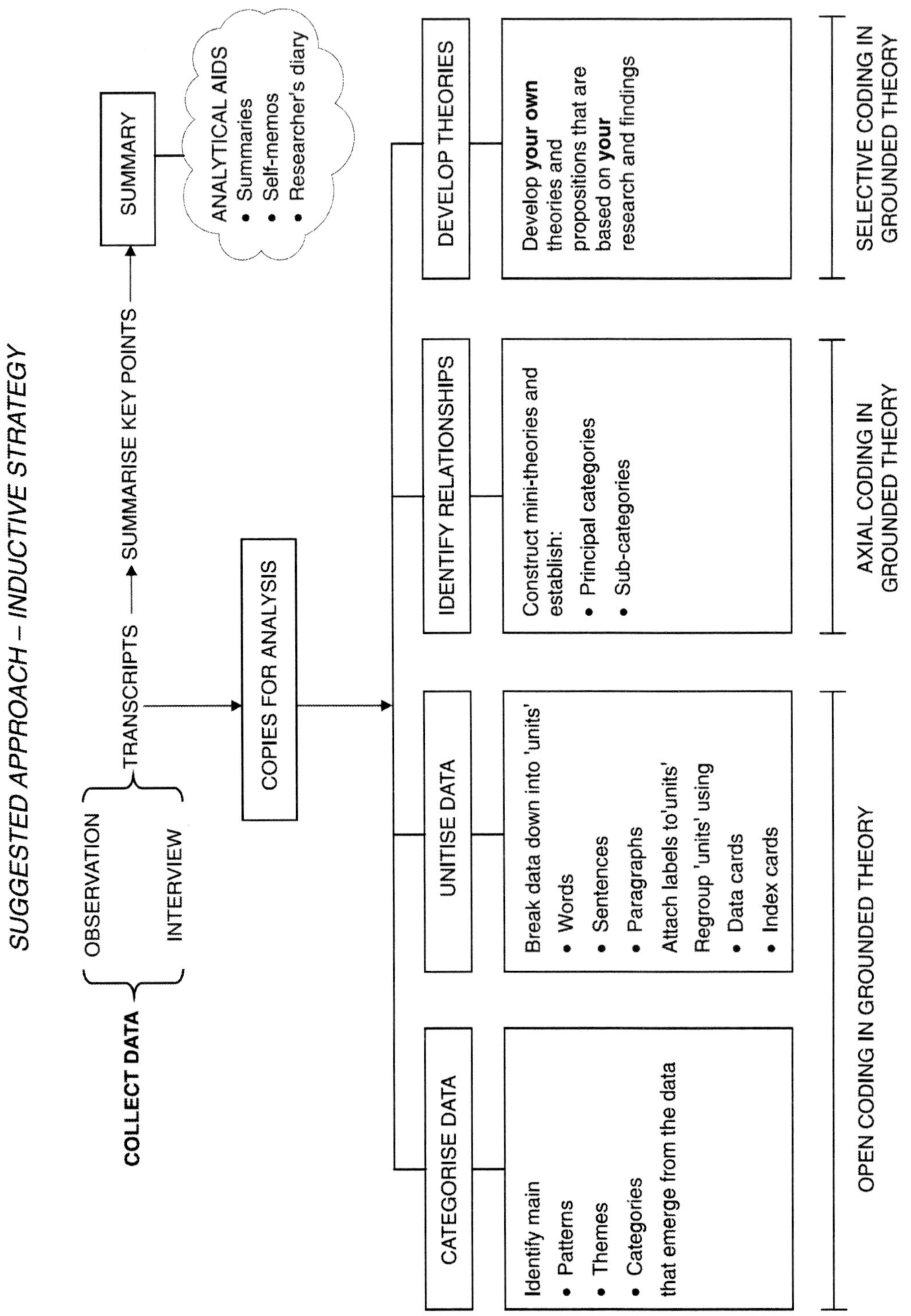

SUGGESTED APPROACH – INDUCTIVE STRATEGY

COLLECT DATA
- OBSERVATION
- INTERVIEW

TRANSCRIPTS → SUMMARISE KEY POINTS → SUMMARY

ANALYTICAL AIDS
- Summaries
- Self-memos
- Researcher's diary

COPIES FOR ANALYSIS

CATEGORISE DATA
Identify main
- Patterns
- Themes
- Categories

that emerge from the data

UNITISE DATA
Break data down into 'units'
- Words
- Sentences
- Paragraphs

Attach labels to 'units'
Regroup 'units' using
- Data cards
- Index cards

IDENTIFY RELATIONSHIPS
Construct mini-theories and establish:
- Principal categories
- Sub-categories

DEVELOP THEORIES
Develop **your own** theories and propositions that are based on **your** research and findings

OPEN CODING IN GROUNDED THEORY

AXIAL CODING IN GROUNDED THEORY

SELECTIVE CODING IN GROUNDED THEORY

Suggested approach – Inductive strategy

4.7 Analytical aids

While analysing your data, you may wish to make a note of any additional information that is relevant to your research. This additional information can be recorded using the following analytical aids.

- Summaries
- Self-memos
- Researcher's diary

Research Methods for Global Marketing Practice

4.7.1 Summaries

Once you have written up your notes, it is a good idea to produce a summary of the key points that have emerged from the data that you have collected. It is also worth making a note of any apparent relationships that you would like to test at a later date in your research. Various other items of information may also be helpful

- Comments relating to the people interviewed or observed.

- Anything unusual or seemingly significant that happened during the interview or observation.

- The setting of the interview or observation.

- Details of any other documents that may be referred to in your analysis, for example, minutes of meetings and internal reports.

4.7.2 Interim summaries

An interim summary may be produced if you wish to summarise your research findings at a certain point. They are a good way of consolidating the information that you have gathered and any initial conclusions that you may have drawn. The information contained in an interim summary includes the following.

- Your findings at a certain point in your research.

- How confident you are in your findings and initial conclusions.

- Ways in which you might be able to improve subsequent data collection and analysis (based on your findings to date).

When you have produced your summary and/or interim summary, it is a good idea to attach a copy of it to your primary research documents (written-up notes or transcripts) so that it can be referred to as you continue with your analysis.

4.7.3 Self-memos

As you progress with your research, you will undoubtedly think of ideas relating to the data collection and analysis techniques that you are using. It is good practice to make an official note of the ideas you have in the form of a self-memo. It is also important to make a note of these notions when you have them, as I'm sure you'll agree, it is all too easy to forget an idea you've had if you don't write it down as soon as possible!

As ideas come to you, you might wish to record them in a reporter's notebook or in the margin of a transcript that you are reviewing. Notes contained within self-memos should be dated and referenced back to the original source documentation. It is also important to update your memos regularly so that as your research progresses, your memorandums are up to date.

4.7.4 Researcher's diary

A researcher's diary is an alternative to a self-memo and is used to record ideas relating to your research project (although these analytical aids are not mutually exclusive, and you may use both methods at the same time). A researcher's diary serves many useful purposes.

- Record ideas about your research.
- Record the ways in which you intend to direct your research.
- Help you to identify categories.
- Help you to identify possible relationships and ways in which they can be tested.

4.8 Use of computers in qualitative data analysis

There are many ways in which qualitative data analysis might be enhanced by computers.

- Small (but significant) pieces of data can be quickly and easily isolated from the mass of information collected during research. Doing it manually could be very tedious and time-consuming.

- Computers are time-saving devices.

- More time and energy can be afforded to analysing data rather than performing routine, mechanical tasks.

- The use of email allows for the speedy transmission of data to anywhere in the world.

- Their flexibility allows new data to be inserted into the appropriate place after data collection.

- Coding. New codes can be added as new categories emerge and data can also be coded in several ways at the same time.

Qualitative data can be analysed using the following software.

- Databases
- Word processing packages
- Computer-assisted qualitative data analysis software

4.8.1 Databases

The large amounts of qualitative data that need to be analysed during your research can easily be stored in a database. It is essential that your database has the following features.

- It is complete
- It is up-to-date
- It allows easy access to all of the documents contained within it

A necessary feature of the software program that you choose is that it includes a comprehensive indexing system which will enable you to access selected items of data quickly and easily.

Most software programs will be capable of searching databases and attaching 'labels' to specific units of qualitative data (text). We can compare this to the (much slower) manual method of reviewing data and marking up the margins with the appropriate labels (codes). It is worth pointing out that the researcher has the advantage of familiarising himself with his research material when he manually codes his data units and that some of this familiarity is lost when computers are used to aid the process of unitising data. (They do, however, alleviate some of the tedium of sorting through the data).

4.8.2 Word processing packages

Word processing packages can also be used to store your research notes and have the advantage of enabling data to be entered directly, such as at the same time interviews or observations are taking place.

Word processors have the following features.

- They enable large volumes of qualitative data (textual) to be easily stored.

- They have 'search' and 'retrieve' options which enable the researcher to locate items of data and group data around the main categories that have been identified.

- Hypertext programs allow units of data to be tagged with invisible markers thus aiding data retrieval and coding.

There are a number of advantages associated with using **computer-assisted qualitative data analysis software** (CAQDAS) when you are faced with large volumes of qualitative data that require analysing. There is however, no statistical package that can replace the interpretation skills of a researcher and so CAQDAS are most usefully employed to carry out the following functions.

- Project management
- Coding of data units
- Retrieval of data units
- Data management
- Testing theories and propositions

Information relating to CAQDAS is constantly changing and can become out-of-date very quickly. Therefore you should visit the websites of some of the most popular CAQDAS in order to obtain the most up-to-date information relating to these programs. Some of these websites will have links on the GMP VLE.

4.8.4 A final word...

If you are considering using computers to aid your qualitative data analysis, you should bear the following points in mind.

- Computers are **not a substitute for the researcher** – and they will need to rely on his/her judgement and skills.

- The researcher will need to **select the computer program** that is **appropriate** for the research being undertaken.

- Sometimes it **may be easier to analyse qualitative data by hand.**

- The skills and experience of the researcher may be lost if too much reliance is placed on computer programs for analysing qualitative data.

1 Describe the nature of qualitative data

- Contextually rich and meaningful data.

2 Evaluate the use of alternative data collection tools in your research project

- The main method of qualitative research is the interview.

- Focus groups concentrate on discussion of chosen topics in an attempt to find out attitudes. They do have limitations despite advantages such as the ability to observe a whole range of responses at the same time.

- Projective techniques attempt to draw out attitudes, opinions and motives by a variety of methods.

- The internet is now a useful means of gathering qualitative data.

- Sampling is appropriate for qualitative research but the concept of generalisability is not applied.

3 Analyse qualitative data

- Contextualisation or content analysis is used for data analysis.

- Quantification, inductive or deductive approaches are taken.

- A general framework has a number of stages, data categorisation, data unitisation, relationship identification and theory development.

1   Recording interviews means that the depth of the content is not lost. Often the pauses, expression used and body language can tell you as much about the response as the actual words used. The choice between video and audio to some extent will depend on how comfortable the respondent is with each alternative. With video it is possible to review the 80% body language following the interview.

2   Research agencies with established online panels would be able to quickly and relatively cheaply form a social network of green consumers to discuss the product and the maximum prices that they would be willing to pay in different scenarios.

3   The need to revise propositions throughout the research cycle means that qualitative research should not be considered a 'quick fix' solution to research problems.

Bell, J., (2005). *Doing Your Research Project: A Guide for First Time Researchers in Education, Health and Social Science*. 4th edition. Buckingham: Open University Press.

Bradley, N., (2007). *Marketing research: tools and techniques*. Oxford: Oxford University Press.

Cooke, N. & Buckley, N., (2008). 'Web 2.0, social networks and the future of market research'. *International Journal of Market Research*. Vol 50, Issue 2.

Collis, J. and Hussey, R., (2003). *Business Research: A Practical Guide for Undergraduate and Postgraduate Students*. 2nd edition. Basingstoke: Macmillan Palgrave.

Denscombe, M., (2003). *The Good Research Guide for Small-Scale Social Research Projects*. 2nd edition. Buckingham: Open University Press.

Dillon, W., Madden, T. & Firtle, N., (1994). *Marketing Research in a Marketing Environment*, 3rd Edition. Illinois: Irwin.

Earls, M., (2003). 'Advertising to the herd: how understanding our true nature challenges the ways we thing about advertising'. *International Journal of Market Research*. Vol 45, Issue 3, pp 311-366.

Easterby-Smith, M., Thorpe, R. and Lowe, A., (2002). *Management Research: An Introduction*. 2nd edition. London: Sage.

Grebenik, E and Moser, C.A., (1962). 'Society: Problems and Methods of Study in Welford'. A.T, Argyle, M, Glass, O. and Morris J.N. (eds)' *Statistical Surveys*. London: Routledge and Kegan Paul.

Miles, M. B. and Huberman, A. M., (2003). *Qualitative Data Analysis: An Expanded Sourcebook*. 3rd edition. London: Sage.

Morgan, D.L., (1997), *Focus Groups as Qualitative Research*. 2nd edition. Newbury Park: Sage.

Palton, M. Q., (1990). *Qualitative Evaluation and Research Methods*. 2nd edition. London: Sage.

Robson, C., (2009). *Real World Research: A Resource for Social Scientists and Practitioner Researchers* Oxford: Blackwell.

Rosenthal, R., (1966). *Experimenter Effects in Behavioural Research*. New York: Appleton Centuary Crafts Strauss, A. and Corbin, J. (1997) (eds) *Grounded Theory in Practice*. Thousand Oaks: Sage

Saunders, M., Lewis, P. and Thornhill, A., (2009). *Research Methods for Business Students*. 5th edition. Harlow: Pearson Education Limited.

Strauss, A. and Corbin, J., (1997). *Grounded Theory in Practice*. Thousand Oaks: Sage

Wass, V. and Wells, P., (1994). 'Research methods in action: an introduction', in Wass, V.J. and Wells, P.E. *Principles in Business and Management Research*. Aldershot: Dartmouth.

Wilson, A., (2006). *Marketing Research An Integrated Approach*. 2nd Edition. Harlow: Prentice Hall.

Yin, R., (2009). *Case Study Research: Design and Methods* (4th ed). Thousand Oaks, CA: Sage Publications.

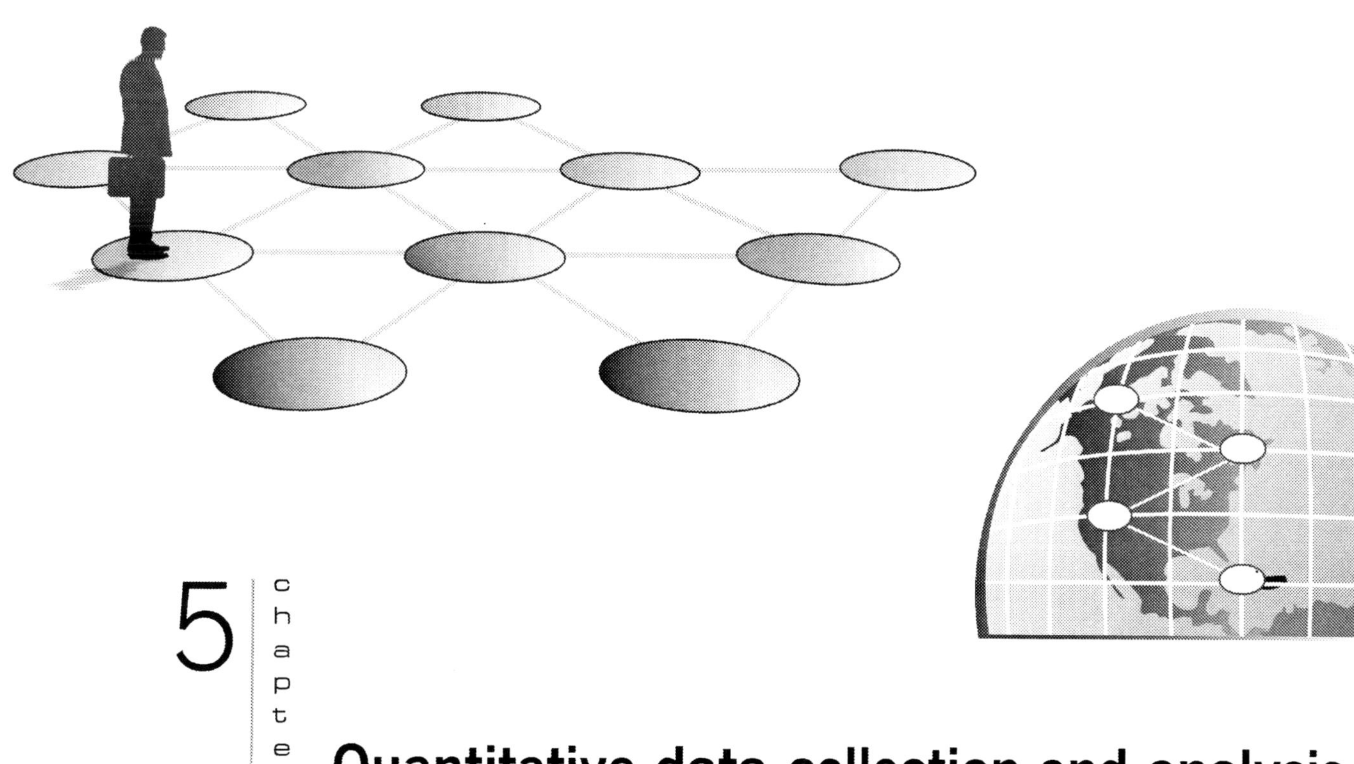

5 | chapter

Quantitative data collection and analysis

The questionnaire is the most widely used means of collecting primary quantitative data, aside from information collected in the ordinary course of business. Observation and experimentation are also commonly used within research designs to collect quantitative data. Both of these approaches are covered in this chapter. It is fair to say however that these methods make extensive use of questionnaires and so we will dedicate time to discussing these. We will cover the use of alternative types of surveys and the questionnaires used in terms of their design, question formats and overall use.

We then move on to look at alternative methods of analysing data. Often (if you are not a statistician) this is a daunting area for students completing a research project. We will show you however that providing you follow the procedure it really is quite straightforward.

This chapter will be supplemented with additional content on the GMP VLE both for this module and when you reach the Research Project for Global Marketing Practice. A series of Fact Sheets will be available to download along with various links to useful external sources of information.

Contents

Types of quantitative data
Data collection tools
Preparation of data for analysis
Graphical techniques
Measures of central tendency and dispersion
Hypothesis testing
Existence of relationships
Existence of trends

In this chapter you will cover the following:

- Distinguish between types of quantitative data
- Evaluate the use of alternative data collection tools in your research project
- Prepare your quantitative research data for analysis
- Identify appropriate graphical techniques for preliminary data review
- Measure central tendency and dispersion
- Identify appropriate statistical techniques to apply within your data analysis
- Establish the existence of relationships in data
- Establish the existence of trends

1 Types of quantitative data

Quantitative data can helpfully be classified into various data types. By doing this we are identifying the level of measurement associated with the variable (Dillon *et al*, 1994).

When we assign numbers to variables (as is the purpose of quantitative data collection) we use these numbers to reflect the correspondence of the number to a specific response. Mathematical operations are then undertaken on the data to establish meanings and research findings.

There are four basic types of measurement levels. These vary in their ability to assign a value, order, rank or simply identify responses. As such the mathematical characteristics they possess also vary. The four data types are:

- Nominal
- Ordinal
- Interval
- Ratio

Any data collected using quantitative methods will fall into one of these measures and so in order to be able to conduct the required level of analysis, it is important to identity the type of data you plan to collect before you start to construct data collection tools such as questionnaires.

1.1 Nominal data

Nominal data is also known as **descriptive or categorical data**. Such data cannot be measured numerically but once the data has been categorised according to its description, it is possible to establish which category occurs the most often and how spread out the data is.

Global case study

A tent manufacturer wishes to identify the favourite colour of tents for their forthcoming new range. They ask the following question to gather responses.

Which is your preferred colour for tents? (Please tick one box).

Red ☐

Blue ☐

Green ☐

1.2 Ordinal data

Ordinal data is also known as **ranked data**. While the actual results of this data are not measured, the overall position of the individual data items is recorded.

Global case study

The tent manufacturer instead asks:

Please rank the following colours according to which you like best.

Red   ____place

Blue   ____place

Green   ____place

With ordinal data you can therefore not only tell which colour is liked best but the order in which different colours are preferred.

1.3 Interval data

Interval data states the differences between two data values. Discrete intervals are identical eg like the markings on a ruler.

Global case study

The tent manufacturer uses the followng scale to ask the question.

Please idenfity for each of the colours shown below how much you like or dislike that colour of tent where 1 indicates you higly dislike the colour and 5 indicates you like the colour.

	Highly dislike	Dislike	Neither like or dislike	Like	Like a lot
Red	1	2	3	4	5
Blue	1	2	3	4	5
Green	1	2	3	4	5

It is possible to tell with this data relatively how much one colour is preferred to another colour. Because there is no absolute zero however you cannot say for example that a score of 4 for red and 2 for green means that red is liked absolutely twice as much as green.

1.4 Ratio data

Ratio data has the same properties as interval data but with an absolute zero.

The tent manufacturer asks:

Please score out of 100 your preference for each colour of tent.

Red ___/100

Blue ___/100

Green ___/100

The following table summarises the mathematical properties associated with each type of data.

Data type	Measurement level	Mathematical property	Example data collection
Nominal	Frequency counts	Small number of tests such as frequency and mode	Which of the following brands do you like best?
Ordinal	Ranking	Wide range of statistical tests which test for order	Rank the following brands in order of your preference
Interval	Relative differences in magnitude	Wide range of statistical tests	Score on a 5 point scale where 1 indicates you like and 5 that you dislike.
Ratio	Absolute differences in magnitude	All statistical tests	How many years have you used brand x?

When we move onto considering statistical methods used to test hypotheses we will return to these levels of measurement. It is important that while designing your data collection tools you bear in mind that this will determine the statistical tests you are able to use.

2 Data collection tools

We mentioned in the introduction that questionnaires are the most commonly employed methods used for quantitative data collection and we have just shown that it is possible to integrate all levels of measurement using this tool.

2.1 Questionnaires

Definition

Collis and Hussey (2003) define **questionnaires** as '*a method for collecting data in which a selected group of participants are asked to complete a written set of questions to find out what they do, think or feel*'. They argue that questionnaires may used within both a positivistic or phenomenological paradigm.

2.1.1 Positivistic paradigm

Collis and Hussey (2003) identify various characteristics relating to the use of questionnaires within a positivistic paradigm.

- They lend themselves to large scale surveys
- Use of closed questions
- Each question can be coded at the design stage

Research Methods for Global Marketing Practice

- Completed forms can be easily analysed with the help of a computer
- Every respondent should be asked, and understand, the questions in the same way

However, Denscombe (2003) has also identified certain weaknesses relating to the use of closed questions.

'There is less scope for respondents to supply answers which reflect the exact facts or true feelings on a topic if the facts or opinions happen to be complicated or not or do not fit into the range of options supplied by the questionnaire.'

'The respondents get frustrated by not being allowed to express their views fully in a way that accounts for any sophistication, intricacy or even inconsistencies in their views.'

The positivistic paradigm is likely to be associated with *explanatory* or *analytical* research where the relationship between variables is examined, usually to try to confirm a cause-effect link.

2.1.2 Phenomenological paradigm

Questionnaires used within a phenomenological paradigm involve certain characteristics.

- Open ended questions are used.
- Coding is done only after the forms have been completed by the respondents.
- They do not lend themselves to large scale surveys.

(Collis and Hussey, 2003)

According to Denscombe (2003) open questions have several advantages.

- The respondent is allowed to decide the wording, length and content of the answer.

- Long answers can easily be accommodated.

- Responses are more likely to reflect the richness and complexity of the views held by the respondents, who are allowed the space to express themselves in their own words.

And there are also potential disadvantages.

- More effort is required of the respondent which might discourage completion of the questionnaire.

- The data obtained is usually 'raw' and may need a lot of time consuming analysis before it can be used.

(Denscombe, 2003)

A phenomenological paradigm is likely to be associated with *descriptive* research, which attempts to describe variances in behaviour.

2.1.3 Role of questionnaires

As indicated earlier in this book, a multi-methods approach is advocated. Hence the use of a questionnaire is likely to be accompanied by some other method of gathering data.

Global case study

A questionnaire is used to identify customer buying habits for say groceries. This research might be supplemented with the use of in depth interviews to explore and understand the underlying reasons and background to the behaviour.

Gill and Johnson (2002) suggest an approach that goes in the opposite direction, where the questionnaire method is used in effect to support an ethnographic strategy.

- Begin data collection with an unstructured and exploratory investigation using overtly ethnographic methods

- Develop theory inductively based on data gathered

- Test the theory using a more structured questionnaire as part of the main study

You might envisage an even more sophisticated scenario than the above that combines observation with a structured questionnaire and an unstructured interview.

Global case study

In a study on the effectiveness of a company's processes for rolling out change management initiatives, the researcher decides to use a multi-methods approach.

- Firstly, the researcher observes employee behaviour and attitudes in implementing the re-engineered processes

- Then a self administered questionnaire is used to identify employees' views on the way the change has been implemented

- Thereafter, unstructured interviews are conducted with management to ascertain their side of the story

---

There is no 'Holy Grail' that will provide a best method of gathering research data. In advocating a multi-methods approach, Denscombe (2003) identifies a series of benefits that might accrue from using different methods to collect data on the same thing.

- Each method can look at the thing from a different angle – from its own distinct perspective – and the researcher, as a means of comparison and contrast can use these perspectives.

- Multi-methods will help the researcher to understand the topic in a more rounded and complete fashion than would be the case had the data been drawn from just one method.

- Alternative methods allows the findings from one method to be checked against the findings from another.

- Seeing things from a different perspective and the opportunity to corroborate findings can enhance the validity of the data. They can help reassure that the findings are not too dependent on the method of data collection.

Denscombe (2003) provides a useful diagram depicting how the data collected from the various methods is *triangulated*.

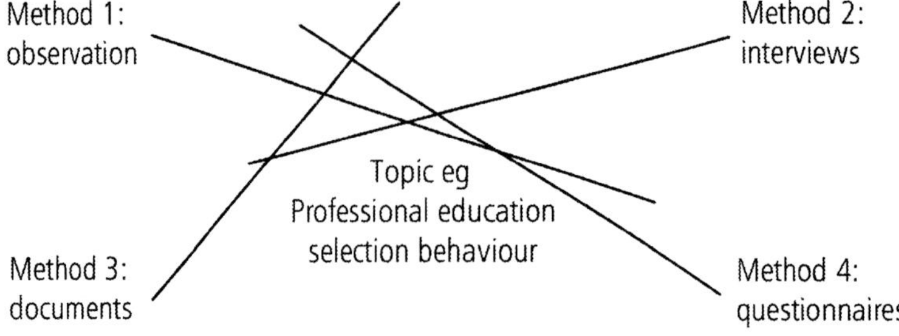

 Research Methods for Global Marketing Practice

Questionnaires are usually divided into two categories:

*   Self administered
*   Interviewer administered

You are likely to have some personal experience of these. For example, you are most likely to have received many questionnaires of the self administered variety through the post and on the other hand you might occasionally have been stopped 'in the street' by an interviewer and asked if you would be willing to answer some questions.

The following diagram reflects the various types of questionnaire that fall within each category.

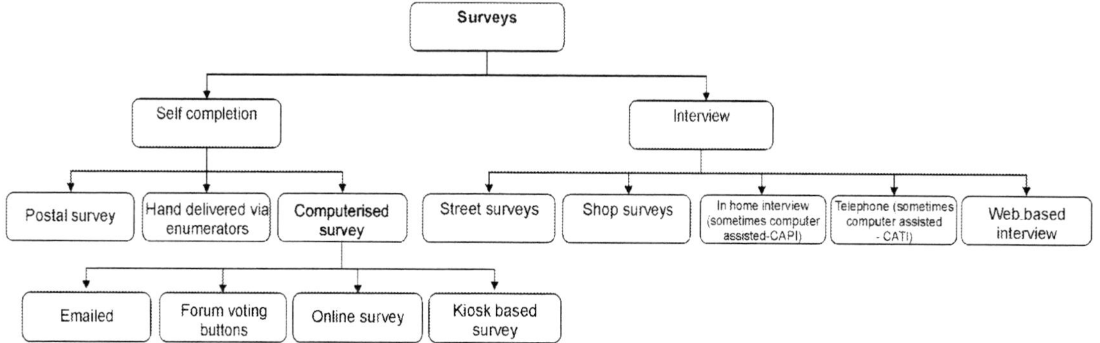

Activity 1

Try to think which is the most appropriate type of survey in your research context and why this might be the case.

Assessment advice

The diagram above shows a large variety of survey types, not all of which will be direcly relevant for your own research. For example forum voting buttons may be too restrictive, kiosk based prohibitively expensive and you are unlikely to be able to access CAPI or CATI facilities. We will therefore not go into detail about these forms.

Whether you decide to use a questionnaire to collect your research data will depend on several factors.

*   The **nature of the information** you need to gather based on your research questions and objectives. The data sought should not be too complex or controversial. Questionnaires lend themselves to *factual* and *opinion* questions.

    For example:

    Factual: What did you eat for dinner last night?
    Opinion: What is your favourite type of food?

*   The **characteristics of the respondents**. They need to amenable and capable of providing meaningful answers. For example, before you decide to use a questionnaire, you need to be aware of the potential respondents' literacy levels when sending out a self-administered questionnaire. They must also have the necessary knowledge and experience.

Similarly, there must be an **open social climate** that encourages respondents to provide full and honest answers.

You need to avoid any loss of *reliability* of your data as result of *contamination* of responses, ie where the respondent makes up, embellishes or guesses at the answers. You also need to be aware that respondents may give unreliable answers for a variety of reasons, such as to please the interviewer, to raise their own self-esteem and to help themselves pass the time of day because they are perhaps lonely!

–   The **necessary sample size**. You need adequate respondents for the questionnaire method to provide meaningful results. You must have some idea of the likely response rate you are likely to achieve.

–   The **amount of data** required. If there is a lot of data to be collected, an interviewer-administered approach may be preferable. Potential respondents may well avoid filling in lengthy questionnaires on their own.

–   **Costs**. You will need to prepare a forecast of the likely costs of administering a questionnaire. These will usually include items such as printing, paper and postage and sometimes travelling and payments to interviewers, if used.

–   **Logistics**. You will also need to draw up a projected timetable of all the various steps that will be involved, including design of the questionnaire, pilot testing, word processing, printing, posting, awaiting replies and follow up of non-respondents.

(Saunders *et al*, 2003 and Denscombe, 2003)

Collection tool	Advantages	Disadvantages
Face to face questionnaire	• Flexibility in terms of mix of questions between open and closed • Can accommodate collection of relatively large amounts of data • Interviewer clarification of questions possible • Interviewer can establish rapport with respondent, read body-language and provide encouragement • Can deal with complex and sensitive questions • Visual aids can be used to help respondent interpret questions • In-depth or 'rich' answers can be obtained • In practice, reasonable response rates are usually obtained	• *Contamination* can creep in as a result of respondents giving unreliable responses or the interviewer distorting the response recorded • Interviewers must be properly briefed and perhaps may need training • The processes may be time consuming and expensive, especially if a large geographical area is being covered
Telephone questionnaire	• Easy to select sample; especially in terms of geographical coverage • In practice, reasonable response rates are usually obtained	• Not everyone has a telephone or is listed in the telephone directory • Complex questions do not work very well

   Research Methods for Global Marketing Practice

Collection tool	Advantages	Disadvantages
Telephone questionnaire (cont'd)	• Flexibility in terms of mix of questions between open and closed • Some personal rapport may be achieved • Cost efficient to administer • Respondent anonymity retained • Responses can be recorded in real time on a computer	• Ethical considerations may be involved, eg catching respondents at inconvenient times • Body language of the respondent cannot be read • Contamination can creep in by respondents giving unreliable responses or the interviewer distorting the response recorded
Postal questionnaire	• Wide geographical reach • Sample selection is facilitated • Respondents can complete in their own time • Much respondent anonymity is retained • Token incentives to complete can be given, eg entry into prize draw, provision of free pen etc. • Optical mark readers can be used to electronically capture responses	• In practice, low response rates usually obtained • The is an element of self-selection in the replies received and hence there may be an element of bias • Clarification of questions cannot be provided • Can be time consuming to administer • Can be expensive in terms of printing and postage costs • Non-respondents should be chased up, usually with a further copy of the questionnaire • Non-respondent bias on the results needs to be borne in mind
Delivery and collection questionnaire	• Similar advantages as postal questionnaires • Response rate can be made higher by the collection process	• Clarification of question cannot be provided • Can be time consuming to administer • Can be expensive in terms of printing and postage costs • Collection can work out to be expensive
Online questionnaire	• Enables large geographically dispersed samples to be administered • Easy to administer, speedy delivery and cost efficient • Input may be automated	• Potential respondents must have a computer and be IT literate • Questions should not be too complex • Response rate is usually moderate to low

(Saunders *et al*, 2009; Collis and Hussey, 2003; Sekaran, 2002)

2.1.6 Dealing with non-responses

You are unlikely to achieve a hundred per cent response rate to your questionnaire. However, a high non-response rate may suggest that the data collected is not representative. This is because there may be material differences between those who respond and those who do not.

Non-responses to a questionnaire may happen for a variety of reasons.

- **Ineligibles**. These refer to people who have been selected, but then do not meet the criteria to participate in the research. This problem should be minimised by being more careful in ensuring that only eligible people are originally included in the population from which the sample is to be drawn.

- **Unavailables**. These may occur for a variety of reasons; the person has gone away, is slow in opening the post, is on holiday or gone away on business. The input of such people will not be represented in the data you collect.

- **Refusals**. These refer to people who have been approached or have received a questionnaire and do not wish to participate in your research, without giving a reason.

Other sources of non-response are:

- There may also be cases where the respondent has agreed to take the questionnaire but has not responded to or refused to respond to certain of the questions.

- Incorrect responses, like ticking more than the required number of boxes. This would also be regarded as a non-response.

Collis and Hussey (2003) cite Wallace and Mellor (1988) regarding three methods for dealing with questionnaire non-response.

- Follow up respondents who do not reply to the first request and try to get responses from them. Compare the data collected from respondents to the first request with the data collected from people who replied to the subsequent follow-up procedures.

- Compare the characteristics of the respondents, with those of the overall population, as far as you know them.

- Compare the characteristics of the respondents, with those of the non-respondents, so far as you can discern them, eg age, gender, occupation etc.

In practice, follow-up procedures are susceptible to the *laws of diminishing marginal returns.* The practical advice is that there are benefits in sending three reminders. This could increase your response rate by as much as a third more respondents.

Bell (2005) suggests that a second request should be sent out a week after the first, but this does seem tight. Taking into account postal delivery times, a ten-day wait before sending out the second request seems to be a reasonable interval. In practical terms, you should enclose a further copy of the questionnaire as well as a stamped envelope for the return mailing.

Zikmund (2002) suggests a more proactive approach whereby you would try to make it more attractive for the prospective respondent to complete a postal questionnaire.

- Enclose a user-friendly covering letter
- Provide an token incentive
- Include strategically placed 'interesting questions'
- Apply a systematic follow-up process
- Use advanced notification of the questionnaire being administered, by telephone or letter
- Sponsorship by a credible body

2.1.7 Self-selection bias

You need to be aware of this problem when using self-administered questionnaires, as it may compromise the representativeness of the responses you collect. Often, the people who take the time to complete a self-administered questionnaire are likely to be those who have strong feelings about the matters covered eg someone who has had a particularly bad or good experience with the organisation.

A hotel provides a customer satisfaction form for guests to complete before they check out. It has 20,000 guests in 2005 and it receives 2,000 customer satisfaction forms from them. 250 responses rated 'food quality' as 'poor' and 500 graded it as 'could be better.' 300 marked the food as 'excellent' and 950 scored it as 'good'.

Would it be valid to conclude that 38% of guests were of the opinion that the hotel's food is poor or could be better, and 62% thought it was good or excellent, considering that only 10% of guests took the trouble to fill-in a form? Is it likely that the forms returned were biased towards those who had an extreme experience with the hotel's food?

Given the low level of responses and the self-administered nature of the questionnaire, there must be some suspicion that the data collected is biased towards the views of those people who developed extreme opinions of the hotel's food.

Generally, with self-administered questionnaires, there is an overrepresentation of people with extreme views and an under representation of people with views in the middle. Given the low level of responses, it would be important to perform some of the follow-up procedures outlined above.

2.1.8 Identifying the data to be collected

The data to be collected and hence the questions you will include in your questionnaire, will be driven by the research questions and research objectives of your investigation.

In keeping with the earlier discussions in this book, the data you gather will also be influenced by whether you are working within a positivistic or a phenomenological paradigm.

- Within a positivistic approach, your questionnaire is likely to be of an analytic or exploratory nature. Your approach will be deductive and attempt to test theory. Variables will be identified and work done to prove relationships between them. (Gill and Johnson, 2002; Saunders *et al*, 2009)

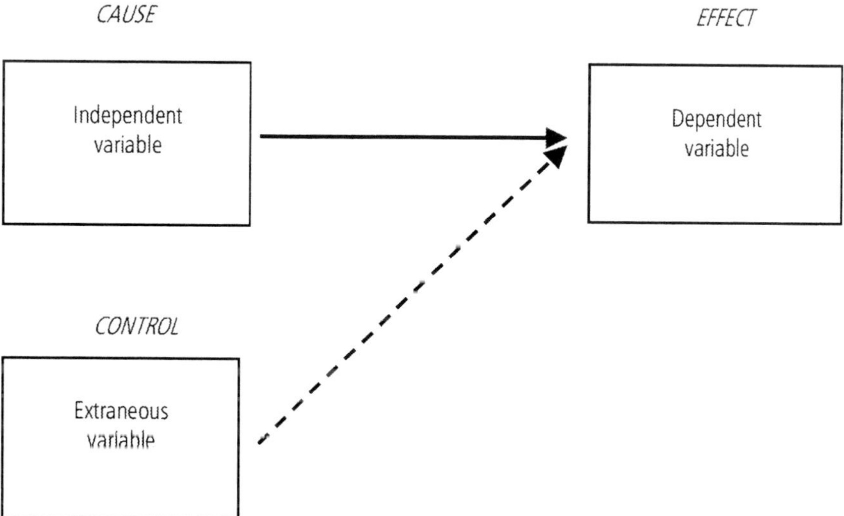

- Under a phenomenological paradigm, your questionnaire is likely to be of a descriptive nature. Your work is likely to be inductive, building theory from the data you collect. You could be studying consumer or employee attitudes and opinions and consumer or organisational behaviour. You will be looking at variability in phenomena. (Saunders *et al*, 2009; Gill and Johnson, 2002)

Within both paradigms, the data accumulated in your literature review will help to inform the nature of the primary data you will need to collect by means of your questionnaire. Saunders *et al* (2009) also suggests that you involve '*colleagues, your project tutor and other interested parties*' in the process of identifying what data is required. They also emphasise that where you are conducting your research within an organisation, it is vital to develop a good understanding of the organisation, including a review of corporate publications and its internet site.

Data requirements table

Saunders *et al* (2009) recommend that you use a *data requirements table* to summarise the data you want to collect. We have reproduced the table from Saunders *et al* (2009) below to provide you with an example of what a *data requirement table* looks like. The table reflects a positivistic approach and would have to be adapted accordingly if you were going down a phenomenological research route.

When identifying the data you need to collect, you will also need to consider the detail at which these are measured. (Saunders *et al*, 2009)

In deciding on the questions to include, you will need to keep in mind how the data collected will be used.

- Plan ahead how the results will be analysed, presented and written up.

- Assess how the data collected might be triangulated with the points identified in your review of literature.

You should be aware of the available techniques that can be used to analyse your data. It would therefore be well worthwhile becoming comfortable with the techniques set out previously before you embark on designing your questionnaire. Otherwise you might fall into one of two traps.

- The data collected is difficult to analyse or interpret.

- There are good data analysis techniques that go unused because you have not collected any relevant data.

You usually only have one opportunity to put you questions to the respondent. Therefore you must make sure you take great care in properly identifying what data is important.

Some years before changes in legislation banned smoking in UK workplaces and public places, a researcher was asked to discover staff attitudes to the possible introduction of a no smoking policy at her workplace. Discussion with senior management and colleagues and reading relevant literature helped her to firm up her objective and investigative questions. A selection of these is included in the extract from her table of data requirements:

Research question/objective: To establish employees' attitudes to the possible introduction of a no smoking policy at their workplace			
Type of research: Predominantly descriptive, although wish to examine differences between employees			
Investigative questions	**Variable(s) required**	**Detail in which data measured**	**Check included in questionnaire ✓**
Do employees feel that they should be able to smoke in their office if they want to as a right? (attitude)	Attitude of employee to smoking in their office as a right	Feel ... should be allowed, should not be allowed, no strong feelings	
Do employees feel that the employer should provide a smoking room for smokers if smoking in offices is banned? (attitude)	Attitude of employee to the provision of a smoking room for smokers	Feel ... very strongly that it should, quite strongly that it should, no strong opinions, quite strongly that it should not, very strongly that it should not	

Investigative questions	Variable(s) required	Detail in which data measured	Check included in questionnaire ✓
Would employees accept a smoking ban at work if the majority of people agreed to it? (behaviour)	Likely behaviour of employee regarding the acceptance of a ban	Would ... accept with no preconditions, accept if a smoking room was provided, not accept without additional conditions (specify conditions), would not accept whatever the conditions	
Do employee attitudes differ depending on: Age? (attribute) Whether or not a smoker? (behaviour)	(Attitude of employee – outlined above) Age of employee Smoker	(Included above) To nearest 5-year band (youngest 16, oldest 65) Non-smoker, smokes but not in office, smokes in office	
How representative are the responses? (attributes)	Age of employee Gender of employee Job	(Included above) Male, female Senior management, management, supervisory, other	

2.1.9 Designing the questionnaire

There is no 'golden formula' for ensuring that you design the perfect questionnaire. You will need to *'apply discretion, make trade-offs and exercise judgment when producing and implementing a questionnaire'* (Denscombe, 2003).

The subject of questions included will cover aspects such as wording, form of question, number of questions and sequencing. Appearance will include aspects such as impression, length, introduction and instructions.

2.1.10 Types of questions

The key characteristics of the various types of questions that could be included in a questionnaire are set out in the following table.

Type of question	Characteristics
Open-ended	• The response is provided in the respondent's own words. – For example, state the three most important attributes you consider when deciding on an exam-training provider? • Useful where there is likely to be a big range of potential responses. • Response is influenced by the amount of space provided to write comments. • The responses need to analysed more carefully before a suitable coding structure can be applied.

Type of question	Characteristics

Closed	• Provide a number of alternative responses from which the respondent is asked to choose, eg:
	
	  – Tick the appropriate box to indicate your gender; male or female?
	    Male ☐ Female ☐
	
	  – Are you working under a training contract? Please tick the appropriate box.
	    Yes ☐ No ☐
	
	  – Would you agree or disagree with the following statement?
	    In today's world, managers require the additional knowledge and skills that qualifications such as an MSc can provide
	    Agree ☐ Disagree ☐
	
	• Minimal writing is required
	
	• Responses can be pre-coded to facilitate subsequent analysis
	
	• The alternatives must be mutually exclusive and collectively exhaustive. If any alternatives overlap, there is likely to be respondent confusion. If the alternatives provided are perceived to be incomplete, the respondents may become frustrated and take the liberty of adding their own alternative.
	
	• Sometimes supplemented at the end of the questionnaire with an open-ended question which gives the respondent an opportunity to comment on anything that has not been adequately covered
	
	• Bell, (2005) identified six types of closed questions
	  – List
	  – Category
	  – Ranking
	  – Scale
	  – Quantity
	  – Grid

List	• A list of responses is provided from which any number may be selected, eg: Please tick the relevant boxes to indicate which of the following Learn-a-lot Limited's products you have used? Textbooks ☐ Revision kits ☐ Revise-a-lot cards ☐ MCQ cards ☐ Learnmore tapes ☐ Learnmore videos ☐ Learnmore CD-ROM ☐ Virtual campus ☐
Category	• The respondent is provided with a set of categories from which one should be selected, eg: How often do you visit the Learn-a-lot website? Tick the appropriate box to indicate your response. Daily ☐ Once a week ☐

Research Methods for Global Marketing Practice

	Once a month ☐
	Once every two months ☐
	Quarterly ☐
Ranking	• The respondent is asked to place things in rank order. This provides data on how these rank in the order of priorities of the respondent, eg:
	Please number the following factors in order of importance in selecting a professional exam-training provider.
	Quality of tutors ☐
	Quality of materials ☐
	Availability of face to face courses ☐
	Catering arrangements ☐
	Teaching methods ☐
	Location of centres ☐
	Historical pass rates achieved ☐
	Leisure facilities ☐
	• In practice, respondents can best deal with seven or eight items to rank in a face to face questionnaire and no more than three to four items over the phone. For face to face questionnaires, it usually helps to show the respondent a prompt card showing the things to be ranked.
Scale	• These types of questions are used to gauge the attitudes and beliefs of a respondent in respect of a specific attribute.
	• The most commonly used form of question is the *Likert scale*, eg:

	Strongly agree	Agree	Disagree	Strongly disagree
Learn-a lot's textbooks are well written	1	2	3	4
The tutors are not supportive	1	2	3	4

• The above format forces the respondent to make a clear choice between agreement and disagreement. You could add an extra alternative by using a five-point scale that provides a "not sure" box in the middle, thereby allowing respondents to 'sit on the fence'.

• Another form of scaling question is the Osgood semantic differential. This type of question takes one attribute and seeks to measure the respondent's views in terms of a series of opposing attributes, eg:

Learn-a-lot's courses are:

Technically accurate					Contain technical errors
Boring					Interesting
Exam focused					Subject focused
Too long					Too short
Student driven					Tutor driven

Type of question	Characteristics					
	Provide an ideal learning environment					Provide a poor learning environment
	Interactive					One way – tutor to student
	Use few teaching techniques					Use many teaching techniques
	Excellent illustrative examples					Poor illustrative examples
	No visual-aids					Many visual-aids
Quantity	<ul><li>These are designed to elicit a numerical response</li><li>They can be used to gather attribute or behaviour data, eg</li></ul>State your year of birth ☐ How long does it take you to get from home to your Learn-a-lot centre, door to door? ☐ minutes					
Grid or matrix	These are usually used for large-scale research, such as national censuses. They are therefore not recommended for academic projects, dissertations, theses and the like.					

2.1.11 Avoiding pitfalls in questionnaire wording

Great care needs to be exercised in the wording of your questions. You must ensure that they are effective in collecting the appropriate data to address your research question and research objectives. The various authorities on questionnaire design give a lot of advice on the wording of questionnaires and the language to use.

Pitfall	Advice
Ambiguity	<ul><li>Be careful that some questions may be interpreted and answered in different ways, eg:</li></ul>When did you decide to study management?<ul><li>This could mean state the date you first made the decision or what event persuaded you to take up management studies.</li></ul>
Imprecision	<ul><li>Beware of using imprecise and vague words like 'on average', 'in the near future', 'a lot', 'relevant' or 'as soon as possible,' eg:</li></ul>How often do you visit relevant websites? A lot ☐ Average frequency ☐ Not much ☐<ul><li>The interpretation of what is a relevant website is likely to differ between respondents, as is their interpretation of 'a lot', 'average' and 'not much'. This will reduce the meaningfulness of the data collected.</li></ul>
Assumption	<ul><li>Be careful of making unwarranted assumptions, eg:</li></ul>What type of computer do you have for the purposes of studies: Desk top ☐ Laptop ☐

| --- | --- |
	• In the above situation, you would need to precede the question with a filter question, eg: Q 4. Tick the appropriate box to indicate whether you have the use of a computer for the purposes of your studies: Yes ☐ No ☐ If your answer is no, skip question 5 and go straight to question 6.
Leading	• These will diminish the quality of you data because the respondent has been led to give a certain answer, eg: • 'Constant reworking of past examination papers is the best way of reinforcing learning. This approach therefore increases the chances of passing an examination.' Tick the appropriate box to indicate whether you agree with this statement. Agree ☐ Disagree ☐ • The above data would probably be better gathered using a multiple-choice question asking respondents to tick the box which they believe represents the most effective approach to preparing for an examination.
Double	• This is what politicians call a two-part question and are usually trained to spot. However, a respondent may be confused by such a question, if there are different responses to each part, but the questionnaire only allows for one answer, eg: Do you enjoy the challenge of attending college and sitting the exams for a qualification? Yes ☐ No ☐ Clearly, a student may enjoy either attending college or sitting exams for a qualification, but maybe not both!
Memory	• Beware of asking questions that rely on the respondent's memory. The older the memories, the less reliable they are likely to be, eg: List the three favourite activities you most enjoyed in primary school.
Negatives and double negatives	• The inclusion of the words 'not' and/or 'no' often confuses respondents and should therefore be avoided, eg: Do you believe it is not a good idea to do no revision before an examination? Yes ☐ No ☐
Hypothetical	• These tend to be of little use and any data gathered will be unreliable, eg: If you won a lot of money, would you carry on working? • There is no way of telling how anybody would react if they won a lot of money, relative to how much they already possessed!

Pitfall	Advice
Sensitive	• Care must be taken to avoid or suitably word sensitive issues • However, this is depends on the current social environment and some consumer surveys include questions about some very personal products.
Socially desirable/prestige	• The reliability of responses to such questions needs to be considered. This may occur for example on questions of personal hygiene such as how many times the respondent baths or showers a week. Hence someone who perhaps baths or showers twice a week may feel it socially necessary to state that he or she does so on a daily basis. Similarly, questions regarding how often the respondent goes on holiday may solicit an inflated response. Hence a person who has occasionally taken a second holiday in any year may state that he or she regularly takes two holidays a year.
Jargon	• Use plain English as much as possible. Avoid technical jargon.

2.1.12 Appearance of questionnaires

The presentation and appearance of your questionnaire is likely to make a positive contribution to its success, especially where it is to be self-administered by the respondent.

This should include a covering letter, introduction, clear but polite instructions and attractive layout. The covering letter should ensure that the research objectives are transparent and provide proper reassurance regarding the context in which the data provided will be used.

Using proprietary questionnaire design and analysis software will enhance the appearance of your questionnaire and help increase your chances of persuading a potential respondent to complete it.

2.1.13 Tips and hints on questionnaire layout

The following advice relates mainly to self-administered questionnaires, but where applicable the underlying principles apply as well to questionnaires administered face-to-face.

Layout aspect	Advice
Easy v difficult questions	• Begin with the easier factual questions and move towards the more demanding or more sensitive questions • If possible, place interesting questions at strategic intervals to sustain respondent motivation in completing the questionnaire • A suitably interesting question at the very end of the questionnaire may encourage the respondent to send it back to you
Instructions	• Ensure that the respondent fully understands how to complete the response. Each question should carry a clear instruction in capitals. INDICATE YOUR GENDER BY TICKING THE APPROPRIATE BOX BELOW. Male ☐ Female ☐
Questionnaire length	• You will need to review the contents of your questionnaire to ensure your document is not too lengthy, without sacrificing the collection of any crucial data.

Research Methods for Global Marketing Practice

Layout aspect	Advice
Location of response boxes	• Practical experience indicates that the response boxes should be positioned as far as possible, in line, towards the right hand edge of the page.
Non-leading sequence	• There is a temptation to design the questionnaire to have a theme running through it whereby the preceding questions virtually lead the respondent to 'inevitable' responses to subsequent questions. You should try as far as possible to avoid this effect, without making your questionnaire seem too disjointed.
Classification questions	• There are differing views as to where questions relating to matters such as the respondent's age, gender, education etc should be best positioned in a questionnaire • Placing such questions towards the beginning of the questionnaire may be beneficial if they are brief and straightforward. However, they may be intimidating and off-putting if they are numerous and or sensitive. Placing them at the end may be a cautious approach.
Quintamensional approach	• This approach was first introduced by Gallup as far back as 1947 and is useful in identifying attitudes and opinions. It involves the following five steps: (i) Apply an open-ended question to identify the respondent's level of knowledge of the topic concerned (ii) Ask a further open-ended question to establish the respondent's general perceptions or feelings about the topic (iii) Administer a closed question to focus on the respondent's attitude towards a specific aspect of the topic (iv) Follow-up with another open-ended question to explore the respondent's justification for their attitudes regarding the specific topic (v) Finally, ask a rating question to identify the strength of the respondent's attitudes towards the specific aspect of the topic being looked at
Use of colour	• If you have the resources, consider using colour in the questionnaire text and with the paper used • Remember that soft warm pastel colours such as pink or cream are more likely to stimulate action than the colder colours such as green or blue. Bright colours like red, purple or orange should be avoided.
Use of tick boxes	• Asking respondents to tick a box is easier for the respondent to understand than asking them to circle alternatives • A completed questionnaire using the tick box approach is also easier to analyse than one using the circle answers approach
Thank you	• The questionnaire should end with a final note of thanks to the respondent • You might also consider providing feedback on your findings. This may help to encourage the respondent to return the form.

Showcards are used to help respondents remember options read to them in face to face surveys. They are used as prompts and can include a list of scale points, phrases to select from, logos to recognise, lists of items, pictorial images or anything else that the researcher needs.

The increased use of computer assisted personal interviewing has meant that the researcher will not necessarily need to use physical cards (similar to children's flashcards) as previously but they can make use of more interactive computerised prompts.

2.1.14 Piloting the questionnaire

You may be tempted to administer you questionnaire as soon as possible after you have designed it. However, the risks of failure could be high. Practical experience indicates that a first draft is likely to contain flaws and problems. The general advice is therefore that you pilot test the questionnaire. Collis and Hussey (2003) suggest that it may take several drafts, with tests at every stage, before you are satisfied that you have got it right.

According to Saunders *et al* (2009) *'the purpose of the pilot test is to refine the questionnaire so that respondents will have no problems in answering the questions and there will be no problems recording the data. In addition, it will enable you to obtain some assessment of the questions' validity and likely reliability of the data collected'*.

In practice, a pilot will also provide you with an opportunity to make sure that your coding system works well and the data can be collated efficiently. Time spent up-front on running a good pilot is likely save you time later in administering the questionnaire and analysing the responses.

From a technical standpoint, a pilot should also be used to provide reassurance on the *content validity* of your instrument ie the data collected is in line with your research question and research objectives.

Ideally the pilot should involve *'a sub-sample of respondents who have characteristics similar to those identifiable in the main sample to be surveyed.'* (Gill and Johnson, 2002)

Saunders *et al* (2009) explain that the number of people you choose for your pilot should be sufficient to include any major variations in your population that you feel are likely to affect responses. They cite Fink (1995) in recommending the minimum number for a pilot as 10.

Bell (2005) suggests that if you cannot do a formal pilot, you should not be averse to having to *'press-gang members of your family or friends'* to provide a modicum of assurance that your questionnaire at least appears to be sensible and that there appear to be no glaring anomalies.

Piloting should provide you with the opportunity to refine and develop the interviewing and social skills you are likely to need and help to highlight any possible sources of interviewer bias.

Bell (2005) sets out the key questions you should ascertain from a pilot.

- How long did the questionnaire take to complete?
- Were the instructions clear?
- Were any of the questions unclear or ambiguous? If so, which, and why?
- Did the participant object to answering any of the questions?
- Was the layout of the questionnaire clear/attractive?
- Any other comments?

2.1.15 Coding the responses

Zikmund (2002) defines coding as *'the process of identifying and classifying each answer with a numerical score or other character symbol.'* This facilitates the transfer of responses from questionnaires to a computer.

So, where you are planning to process your data by computer you will have to code up the questionnaire responses. With closed questions this would be done preferably before you administer the questionnaire.

With open-ended questions, you would need to have a look at the responses you have received, classify these into suitable categories and then decide on a coding structure.

The following example is adapted from a UK Rail Traveller's Survey issued by the Strategic Rail Authority (before it was disbanded in 2006).

Q12a **How much inconvenience, if any, have you experienced over the last 6 months due to the disruption of rail services in general?**

A lot of inconvenience	A little inconvenience	Not much inconvenience	No inconvenience	
1	2	3	4	(401)

Q12b **Please write in the space provided below, in your own words, what your experiences have been of the services over the last six months. Please summarise any disruption that you have experienced personally.**

....

....   (402)

....

2.1.16 Testing for reliability

Easterby-Smith *et al* (2002) explains reliability as follows:

- Under a positivist approach, reliability refers to whether your questionnaire will yield the same results on different occasions, assuming that there has been no real change in what is to be measured.

- Under a phenomenological approach, reliability refers to whether similar observations will be made by different researchers on different occasions.

Bell (2005) explains the concept using the example of a watch. If every time you look at it, it is consistently 10 minutes fast, it could be useful because you can adjust for the ten minutes. However, if it could be anywhere between say 5 to 15 minutes fast you would not be able to rely on it.

Collis and Hussey (2003) explain that for positivistic research work, reliability has to be high and you must be able to *replicate* your results.

For a phenomenological study, reliability may not be given so much status, or it may be interpreted in a different way. Here we are concerned with whether similar observations and interpretations can be made on different occasions and/or by different observers (Collis and Hussey, 2003).

The data you collect might not be reliable for a variety of reasons relating to the questionnaire.

- The questions may contain errors, eg the wording may be ambiguous.
- The respondent may become bored.

(Collis and Hussey, 2003)

Sekaran (2002) distinguishes reliability between *stability* and *consistency*.

Stability refers to the ability of a questionnaire to maintain stability over time, whatever the testing conditions and the state of the respondents themselves. Two tests of stability are the *test-retest method* and the *parallel form test method*.

Consistency indicates the *homogeneity* of the questionnaire or sub sections of the questionnaire. It looks at whether the items and subsets of items in the questionnaire are highly correlated. Two ways of identifying consistency are the *interim consistency method* and the *split halves method*.

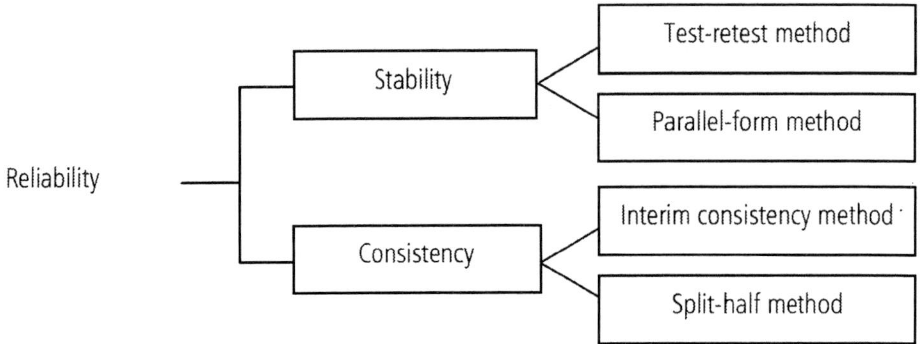

Test-retest method

- The same questions are administered to the same group of respondents at two different dates.

- The time interval must not be too long that it becomes a factor itself in any variation in responses.

- The test-retest co-efficient is computed to provide an index of reliability.

- The practical difficulty is being able to persuade people to complete the same questions twice and in a conscious way.

Parallel form method

- This involves asking the same question in a different way, eg:

    - Rice satisfies my hunger more than pasta? Yes/No

    - Pasta satisfies my hunger more than rice? Yes/No

- The obvious problem here is making the questions look different without introducing material differences in their impact.

- The questions would also appear in different locations within the alternative questionnaires.

- The results are correlated to determine the level of error arising from wording and ordering of questions.

Interim consistency method

- Every response is correlated with every other response in the questionnaire and a correlation co-efficient, usually Chronbach's alpha. (Sekaran, 2002)

- The downside of the method is that it involves substantial computing facilities and software

Split-halves method

- The responses are split into two equal halves, eg into those relating to odd numbered questions and those relating to even numbered questions.

- A split-halves co-efficient is then computed.

Generally, these methods involve a lot of work and you should talk to your supervisor regarding what you need to do to ensure the reliability of your questionnaire.

For background reading in this area you might have a look at Collis and Hussey (2003).

3 Preparation of data for analysis

Once you have collected your data using one of the methods described in the previous chapter, you will need to prepare your data for analysis. The preparation of data for analysis involves the following.

- Data editing and data cleaning
- Insertion of data into a data matrix
- Data coding
- Weighting of cases: differential response rates

3.1 Data editing and data cleaning

The main objective of data cleaning is to *'identify omissions, ambiguities, and errors in the responses'* (Diamantopoulos and Schlegelmilch, 2000).

The process of data cleaning is often referred to as data editing. Data editing can take place either when the data is being collected or immediately after it has been collected. Data editing during data collection is known as the 'field edit'.

3.1.1 'Central office edit'

Once the data has been collected, an edit at central office deals with identifying any of the data which is ambiguous, inconsistent or missing. It is important that any problems with the data are identified before attempting to analyse it as failing to do so will result in unnecessary delays in obtaining the results of the data collected.

The main weakness of a 'central office edit' is that it is unlikely that you will be able to clarify any ambiguous, missing or inconsistent data with the original respondent. However, there are ways that you can minimise these problems from occurring. For example the likelihood of obtaining ambiguous answers might be minimised by ensuring that your questions do not contain any elements of ambiguity.

Global case study

If you are asking respondents when they last went on holiday, make sure that you state whether you would like them to give their answer in terms of:

- Weeks
- Months
- Years
- Actual dates (ie 24/09/2000)

Imagine the extra work that you could be creating for yourself if you had to convert all of the responses given into the same units (weeks, months or years)!

3.1.2 Missing data

One of the most common problems that you will come across when editing your data is that of missing data. Respondents may not answer a question for one of the following reasons.

- The question was not applicable to the respondent
- The respondent refused to answer the question
- The respondent was unable to answer the question
- The respondent omitted the question (in error)

Depending on the data you are collecting, you will need to decide how best to deal with questions that respondents have not answered. Sometimes it will be necessary to identify which one of the four reasons above applies and sometimes you will feel that such identification is unnecessary. You should not fabricate answers to omissions. You may well be able to discuss or explain the omission of responses when you are writing up your results. For example, if respondents consistently fail to give a response to a specific question you might try to offer some reasoning for the pattern of non-response.

3.2 Insertion of data into a data matrix

The best way of presenting raw data (after it has been cleaned and edited) is in the form of a table that is known as a data matrix. Most analysis software will only accept data that has been entered into a data matrix.

Sometimes primary data can be entered and saved on a computer file when it is collected. Examples of such data collection methods are as follows.

- Computer-aided personal interviewing (CAPI)
- Computer-aided telephone interviewing (CATI)
- Online questionnaires

It is also worth noting that secondary data which is collected from the internet or from CD-ROMs may not need to be re-entered into a data matrix if it is saved in an appropriate format.

If your data is not entered into a computer at the time of collection, or if it is not saved in an appropriate format, you will need to enter it into a data matrix.

3.2.1 Data matrix

The data matrix is the **starting point for analysis** and its structure defines the extent to which data can be analysed. A simple data matrix has a number of **rows** which are **horizontal** (say, N) and a number of **columns** which are **vertical** (say, M).

In order to demonstrate many of the aspects of quantitative data analysis, we are going to refer to the results of a survey that was conducted in order to determine information about the following variables for 1,000 respondents.

- Age
- Sex
- Nationality
- Annual income
- Number of holidays last year
- Preferred holiday type
- Whether they ever holiday in Britain
- If they holiday in Britain, their preferred accommodation

Figure 1 Variables

When we wish to refer to this survey in this chapter, it will be called the 'holiday survey'. Information relating to three of these respondents is shown in the extract of a simple data matrix, Figure 2 below.

COLUMNS = M							
M1	M2	M3	M4	M5	M6	M7	M8
Age	Sex	Nationality	Annual Income	Number of holidays last year	Preferred holiday type*	Holiday in Britain?	If M7 = yes Preferred accommodation? **

ROWS = N		M1	M2	M3	M4	M5	M6	M7	M8
	N1	37	Female	British	£46,000	3	Beach	Yes	B&B
	N2	39	Male	Asian	£98,000	1	Skiing	Yes	Hotels
	N3	45	Male	Australian	£42,000	0	Camping	No	N/A

*   Respondent to choose from one of four options (beach, skiing, camping, other)

**   Respondent to choose from one of four options (B&B, hotels, self-catering cottages, other)

Figure 2 Extract from a simple data matrix

You will see in the extract above that the rows (N1 – N3) indicate the number of units which are being studied ie the size of the sample (or population under review). The columns (M1 – M8) indicate the number of variables which are being investigated. In this case, the extract from the data matrix contains information (variables) about three individuals (N1, N2 and N3). The information relates to how much they earn, what sort of holidays they like to take and so on. You can see that this data can be summarised very neatly in the form of a data matrix. We will return to this example when we look at **data coding** in the next section of this chapter.

3.3 Data coding

3.3.1 Code books

A code is a system of numbers or letters which are used to represent others. In quantitative data analysis you are most likely to use a system of numerical codes which can be read by a computer. A list of codes which gives a detailed explanation of responses is known as a code book.

The code book relating to the 'holiday survey' is shown below.

Code	Description	Variable
1	18-25 years	
2	26-35 years	
3	36-45 years	
4	46-55 years	Variable M1 = AGE
5	56-65 years	
6	More than 65 years	
10	Male	Variable M2 = GENDER (see Note 1)
11	Female	
20	British	
21	American	
22	European (excluding British)	
23	African	Variable M3 = NATIONALITY (see Note 2)
24	Asian	
25	Australian/New Zealand	

Code	Description	Variable
30	Up to £10,000	Variable M4 = ANNUAL INCOME (see Note 3)
31	£10,001 - £20,000	
32	£20,001 - £30,000	
33	£30,001 - £40,000	
34	£40,001 - £50,000	
35	£50,001 - £60,000	
36	More than £60,000	
40	Beach	Variable M6 = PREFERRED HOLIDAY TYPE
41	Skiing	
42	Camping	
43	Other	
50	YES	Variable M7 = HOLIDAY IN BRITAIN?
51	NO	
60	B&B	Variable M8 = PREFERRED HOLIDAY ACCOMMODATION
61	Hotels	
62	Self-catering cottages	
63	Other	
64	N/A – don't holiday in Britain	

Figure 3 'Holiday survey' codebook

Note 1. This is descriptive data – you are either male or female, and each gender is coded 10 or 11.

Note 2. This is descriptive data – in most cases respondents will be described as having one main nationality.

Note 3. When entering this information into the data matrix, initially the salary is coded as the actual value. For example, £46,000 would be coded as 46000 as shown in Figure 4 below.

Once this data has been entered into the matrix, you can use analysis software to re-code the data by identifying the data as belonging to one of the categories coded 30-36.

Note 4. Note that there is no code for the number of holidays taken last year. The number of holidays must be stated to the nearest whole number (discrete data).

We can use the 'holiday survey' codebook in order to code the data in Figure 3 and produce the table below.

		COLUMNS = M							
		M1	M2	M3	M4	M5	M6	M7	M8
		Age	*Sex*	*Nationality*	*Annual Income*	*Number of holidays last year*	*Preferred holiday type*	*Holiday in Britain?*	*If M7 = yes Preferred accommodation?*
ROWS = N	N1	3	2	1	46000	3	1	1	1
	N2	3	1	5	98000	1	2	1	2
	N3	3	1	1	42000	0	2	2	5

Figure 4 Extract of a coded data matrix

3.3.2 Re-coding

Sometimes actual values may be used as codes for quantifiable data. For example, Column M4 in the coded data matrix (Figure 4) above includes salaries as actual values: £98,000 is originally coded as 98000. Analysis software can be used to re-code data into categories which contain similar values. Figure 5 below shows how annual incomes (for respondents N1, N2 and N3) can be re-coded using the 'holiday survey' codebook.

Original Annual Income	Coded Annual income	Re-coded Annual Income
£46,000	46000	5
£98,000	98000	7
£42,000	42000	5

Figure 5 Re-coding of annual incomes using 'holiday survey' code book

3.3.3 Existing code books

If you are using secondary data, you may find that a coding scheme is already in existence. Such a scheme is known as an **existing code book** and would have been created when the data was first collected. However, there will be circumstances when you will need to create your own code book, for example when you are collecting primary data. (You may also need to create your own code book in certain situations where you are using secondary data. You will need to look at each case individually and consider the level of precision that will be required).

Existing code books can be used for many variables, such as the following which were identified by Saunders *et al* (2009).

- **Industrial Classification** (Source: Central Statistical Office 1997)
- **Occupation** (Source: Office of Population Censuses and Surveys, 1990a, b)
- **Social Class** (Source: Office of Population Censuses and Surveys, 1991)

Using existing code books affords you various benefits.

- They save time

- They have usually been extensively tested

- You can make comparisons with other survey results because you will be using a common language (ie the same code book)

The codes contained within the existing code book are known as **pre-set codes** and should be included on your survey forms so that they can be referred to by the person filling in the form.

3.3.4 When should you code your data?

You can either code your data at the **time of collection** or **after collection** – it is up to you. However, various considerations are likely to influence your decision.

- If you are confident that there will be **limited number of responses** to your questions, then you will probably have a list of **pre-set codes** and can code your forms as you collect the data

- If you are unsure of the number of likely responses to your questions or if there are a large number of possible responses, then it is best to create your code book once you have reviewed a sample of responses. In such a situation, it will be best to code your responses **after** data collection.

3.3.5 Creating a code book

If you are not in a position to use an existing code book, you will need to bear the following points in mind when you create a code book.

- Examine the preliminary results from the first batch of questionnaires completed.

- Draw up a list of possible groups.

- Consider the level of precision required and then create further sub-groups using this information.

- Allocate codes to the groups and sub-groups that you have identified. Make sure that groups which may be combined into larger groups (made up of several similar groups) are given adjacent codes so that the data may be re-coded easily.

If you refer to the 'holiday survey' code book in Figure 3 you will see that the codes for variable M1 start at 1 and end at 6. Codes for variable M2, however, do not start until 10. This allows additional codes (7, 8 and 9) to be inserted if the age categories need to be amended at some point in the future.

3.3.6 Missing data codes

We have already mentioned that there are four main reasons why respondents may not respond to a question and therefore data might be missing from the data matrix. Missing data is usually indicated by a **missing data code** and is excluded from analysis if necessary. Sometimes, however, it might be necessary to indicate the reason why the data was missing by assigning different **missing data codes**.

3.3.7 Multiple response and multiple dichotomy methods of coding

You may come across situations where there are many possible answers to a question asked in a survey. For example, suppose the 'holiday survey' asked one further question, such as:

'List the things you like about going on holiday.'

Respondents are likely to have provided various possible answers.

- Sunbathing
- Sightseeing
- Relaxation
- Socialising

The multiple dichotomy method of coding would involve identifying whether each respondent either 'listed' or 'did not list' each of the variables (responses) shown above. A separate code would then be allocated to 'listed' and 'did not list' (for example, 1 and 2 respectively). This would then enable you to identify how many respondents listed 'not going to work' or 'eating out', for example, things they like about going on holiday.

An alternative to this coding method is the multiple response method. While conducting your survey, you might wish to enquire (once again) as to the reasons why people like going on holiday. The multiple response method of coding would ask the following question:

'Which of the following reasons for enjoying holidays would you consider to apply to you?'

- Sunbathing
- Sightseeing
- Relaxation
- Socialising

Possible codes would be:

    1 = applies
    2 = does not apply

Research Methods for Global Marketing Practice

Therefore each reason would be coded 1 or 2 as follows.

Variable	Response
Sunbathing	1
Sightseeing	2
Relaxation	1
Socialising	1

The main difference between these coding methods is that the **multiple dichotomy method** asks an open question, and the number of respondents listing the same variables (reasons) can then be determined. On the other hand, however, the **multiple response method** asks a closed question, and each respondent must then select which variables (reasons) apply to him and which do not.

3.3.8 Coding errors

Errors are likely to occur no matter how carefully you code and enter your data. It is therefore extremely important that you check your data for errors, even if this is a very time-consuming process. If data is not checked, the final results may be incorrect and you will be in danger of drawing incorrect conclusions from your results. Remember the following points when you check your data for errors.

- Make sure that there are no illegitimate codes. For example, with numerical codes, make sure that the number 0 (zero) is not replaced by the letters o or O. Similarly, the letter I may be incorrectly used instead of the number 1.

- Investigate any illogical relationships. For example, if a respondent has an annual income of £5,500 and took seven holidays last year, this might require further investigation to ensure that the details have been entered into the matrix correctly. Perhaps the annual salary on the original data sheet was £55,000. You will need to check the coding and the original source material.

- Ensure that certain rules in filter questions are followed correctly. For example, in the coded data matrix extract (Figure 4) above, in row N3, the respondent has said that he doesn't holiday in Britain (code = 2) and it therefore follows that the response to variable M8 should be coded 5 (N/A – never holiday in Britain). If the response to variable M8 were coded 1, 2, 3 or 4, a coding error would have occurred. Statistical analysis programs should be able to pick up errors such as these by running a series of checks on the program.

3.4 Weighting of cases: differential response rates

In some circumstances it may be necessary to apply different weights to your data, for example if you have used stratified random sampling to obtain your sample, and have obtained different response rates for each stratum. The following example will show you how to weight your data if you have differential response rates when selecting a sample using stratified random sampling.

Global case study

Let us suppose that you select a sample for a survey using stratified random sampling and that there are three strata, A, B and C. The response rates for each stratum are as follows

Stratum	Response rate
A	90%
B	75%
C	60%

The weights to be applied to each stratum are as follows.

Stratum A = 90/90 = 1.0
Stratum B = 90/75 = 1.2
Stratum C = 90/60 = 1.5

Note that the weight for the stratum with the highest response rate will always be 1.

Most statistical analysis software will apply weights to your data if the weights are entered as separate variables in your data matrix. The easiest way to calculate weights is to compare the response rates of each stratum with the response rate of the stratum which has the highest response rate (Stratum A in the above example).

When you have **coded** and **entered** your data into your **data matrix** you will be ready to start your **analysis**. In the next section, we will be looking at ways in which different types of graph can help you to understand more clearly the data you have collected. We shall start by looking at how graphs can be used to show individual results.

4   Graphical techniques

The first step once all data is entered and checked is to gain a flavour of the results and prepare a graphical overview. This is highly illuminating and helps to identify any anomalies in the data.

4.1 Presenting individual results

There are a number of ways in which graphs can be used to present individual results.

- **Frequency distributions** are a way of presenting specific values
- **Bar charts** and **histograms** show highest and lowest values
- **Line graphs** can demonstrate trends.
- **Pie charts** can highlight proportions
- **Frequency polygons, bar charts, histograms and box plots** can reflect distributions

We shall be looking at each of these methods of data presentation in turn.

4.1.1 Frequency distributions

Data sets are generally made up of a limited number of data values. Frequency distributions are simply tables which record the number of times (frequency) that a value or characteristic occurs.

If there is a large set of data or if there are many different data groups, it is often convenient to group frequencies together into classes. This is known as a grouped frequency distribution.

We shall be using frequency distributions extensively in this chapter in order to demonstrate the range of graphical techniques that are available for presenting your data.

The following frequency distribution has been extracted from the results which were obtained from the 'holiday survey' when 1,000 respondents were questioned about their age, annual salaries and holiday preferences. We looked at some of the results of this survey when we demonstrated how data matrices are prepared and how results can be coded.

Age (to the nearest year)	Code	Frequency
18-25	1	120
26-35	2	220
36-45	3	320
46-55	4	190
56-65	5	120
more than 65	6	30
	Total	1,000

Figure 6 Frequency distribution showing ages of 'holiday survey' respondents

The frequency distribution in Figure 6 above shows the range of different ages of the 1,000 people who were surveyed. This distribution is made up of discrete data (ie data which can only take on whole numbers – respondents have been asked to state their age to the nearest year). It also demonstrates how the different ages of the respondents have been grouped together in order to produce a grouped frequency distribution.

Figure 7 below shows a grouped frequency distribution which is made up of **continuous data.** It shows how many of the people surveyed earn salaries that fall within certain bands of income.

Annual Income £	Code	Frequency
10,000 or less	1	105
10,001-20,000	2	370
20,001-30,000	3	290
30,001-40,000	4	180
40,001-50,000	5	40
50,001-60,000	6	10
more than 60,000	7	5
	Total	1,000

Figure 7 Frequency distribution showing annual incomes of 'holiday survey' respondents

4.1.2 Bar charts

Bar charts are one of the most common methods of presenting data in a visual form. They are simply charts in which quantities are shown in the form of bars. There are three main types of bar chart.

- Simple bar charts
- Component bar charts, including percentage component bar charts
- Multiple or compound bar charts

Simple bar charts

The results from our 'holiday survey' revealed that the following responses were obtained in relation to Variable M8 (preferred accommodation of those who holiday in Britain).

Preferred accommodation	Code	Frequency
B&B	1	50
Hotel	2	290
Self-catering cottages	3	350
Other	4	130
n/a – don't holiday in Britain	5	180
	Total	1,000

Figure 8 Frequency distribution showing preferred accommodation of respondents who holiday in Britain

The results shown in Figure 8 above could be shown graphically in the form of a simple bar chart. Notice that the data that we are presenting here is **categorical (nominal) data** since it cannot be measured. It is

also evident from the graph that self-catering holidays are the most popular while B&B holidays are the least popular.

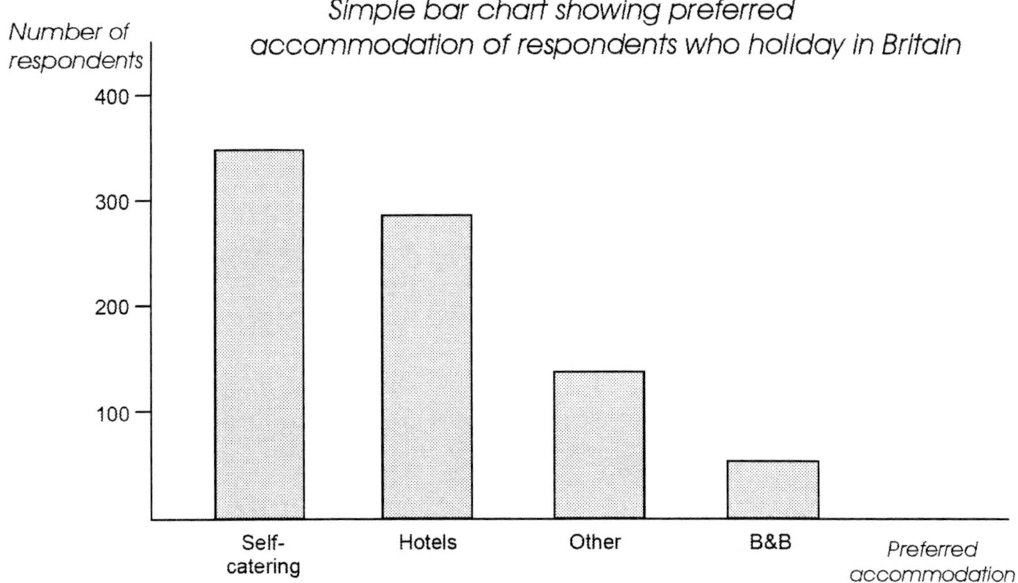

Figure 9 Simple bar chart

Note that the results of the 180 respondents who do not holiday in Britain are not included on the graph.

4.1.3 Histograms

A histogram is a chart that looks like a bar chart except that the bars are joined together. On a histogram, frequencies are represented by the areas covered by the bars, and not the heights of the bar (as is the case with bar charts).

Histograms are generally used for continuous data, and the continuous nature of the data is indicated by the way in which the bars do not have any gaps between them.

The frequency distribution in Figure 7 above could be presented as the following histogram.

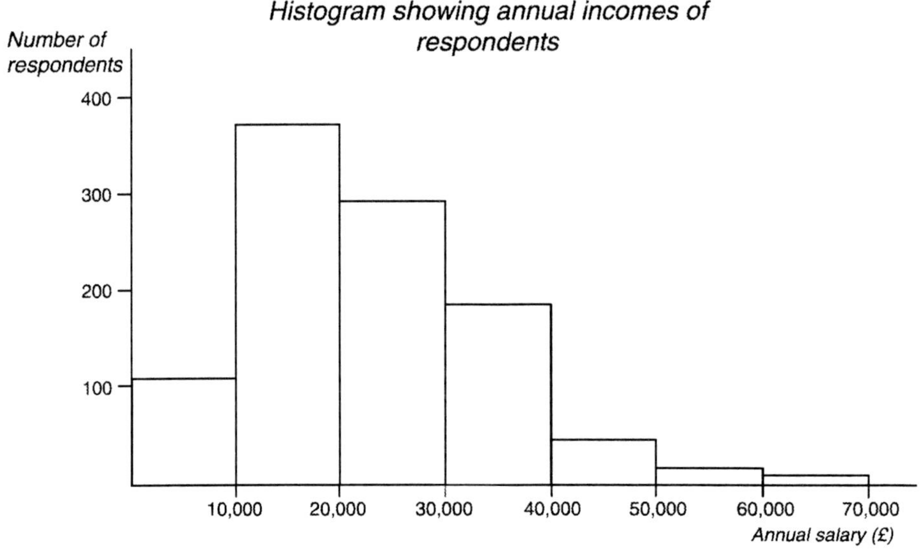

Figure 10 Histogram

The graph above has equal-width class intervals and so the highest and lowest values are easy to identify (for example, the highest value is the salary band £10,001-£20,000). The heights of the bars still represent the frequencies when the class widths are equal.

When class widths are unequal, it is the area of the bar which represents the frequencies and not the height. Analysis software is not able to cope easily with drawing histograms with unequal-width classes and therefore treats such histograms as a variation of a bar chart.

Histograms are not a particularly good way of presenting a distribution, they are far better presented by frequency polygons. Frequency polygons can be drawn from a histogram by joining up the mid-points of the histogram bars. The histogram in Figure 10 could be converted into a frequency polygon as follows.

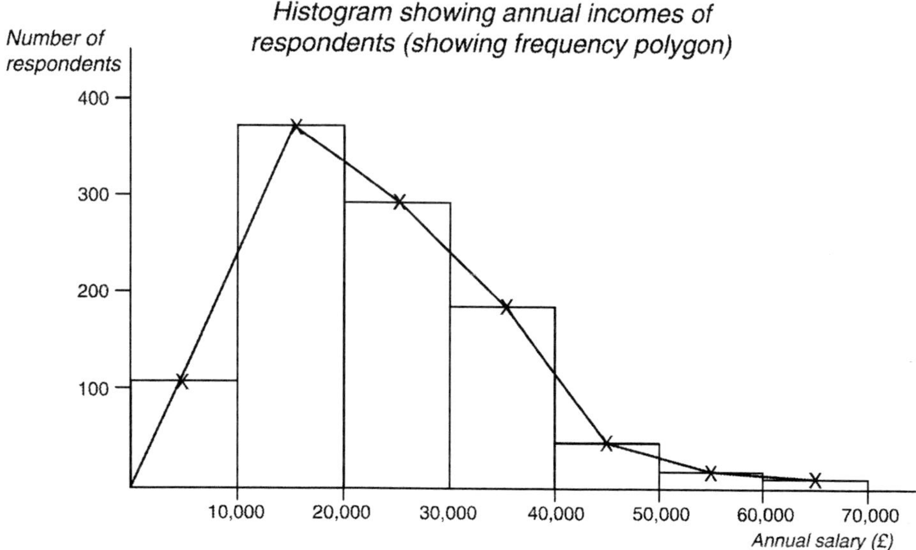

Figure 11 Frequency polygon

Note that most analysis software will treat a frequency polygon as a version of a line graph. We shall be looking at line graphs in the next section.

4.1.4 Line graphs

Line graphs are the most suitable diagrams to draw if you wish to show the trends of individual results.

If we wanted to compare the average number of holidays taken per year by the general public over the last ten years, as per our 'holiday survey,' we could plot our results on a line graph such as the one shown in Figure 12 below.

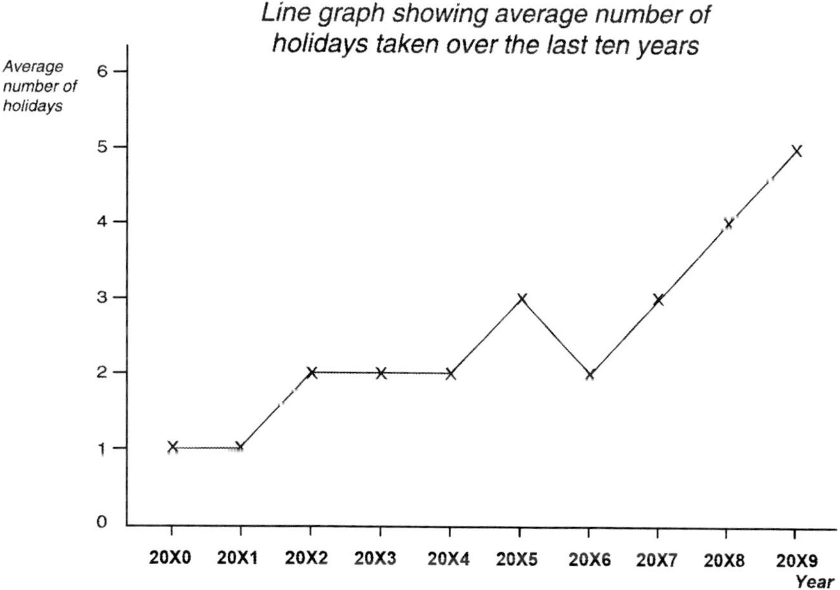

Figure 12 Line graph

It is evident from the line graph that the general trend in the number of holidays taken by respondents is increasing over time.

A pie chart is a chart that is used to show pictorially the relative size of component elements of a total.

If we wished to show the types of holiday preferred by the respondents in our 'holiday survey', a pie chart would be able to present this data very clearly and would show the proportion of respondents that had a preference for each type of holiday. Suppose our 'holiday survey' revealed the following results.

Preferred holiday type	Code	Frequency	Percentage
Beach	1	350	35%
Skiing	2	290	29%
Camping	3	280	28%
Other	4	80	8%
Total		1,000	100%

Analysis software could present this data as follows.

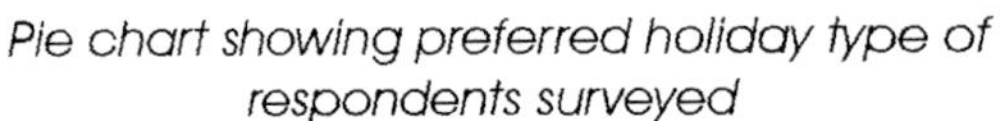

Pie chart showing preferred holiday type of respondents surveyed

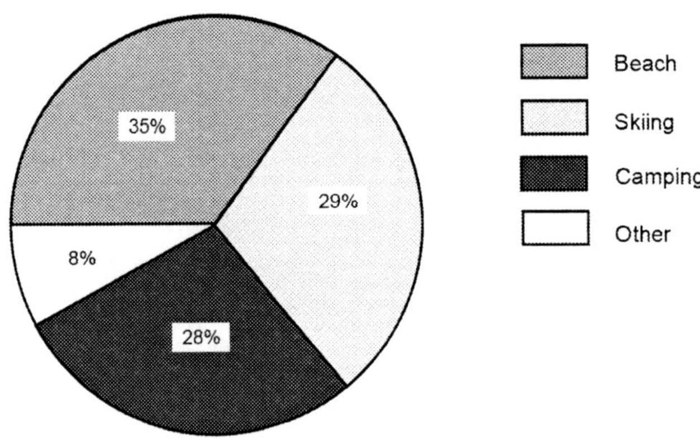

Figure 13 Pie chart

Note that a pie chart is a circle and that a circle has 360 degrees. The pie chart is therefore drawn by allocating 35% of the 360 degrees (126 degrees) to beach holidays, 29% of the 360 degrees (104 degrees) to skiing holidays and so on. Don't worry - the data analysis software that you are using will draw your pie charts for you!

4.1.6 Box plots

We have already said that distributions are shown most easily on frequency polygons, bar charts, histograms and box plots. We have looked at all of these data presentation methods except for box plots.

Box plots are often included in more advanced statistical analysis software programs. A box plot is a diagram that displays the distribution of data pictorially and also identifies the following statistical values.

- Median (middle value of a distribution)
- The range of the data
- Extreme values (highest and lowest values in the range of data)
- Inter-quartile range (the middle 50% of the distribution)

Don't worry about these terms – we shall be looking at them in more detail when we study the statistical techniques that you need to be familiar with. The annotated sketch below identifies the main features of a box plot.

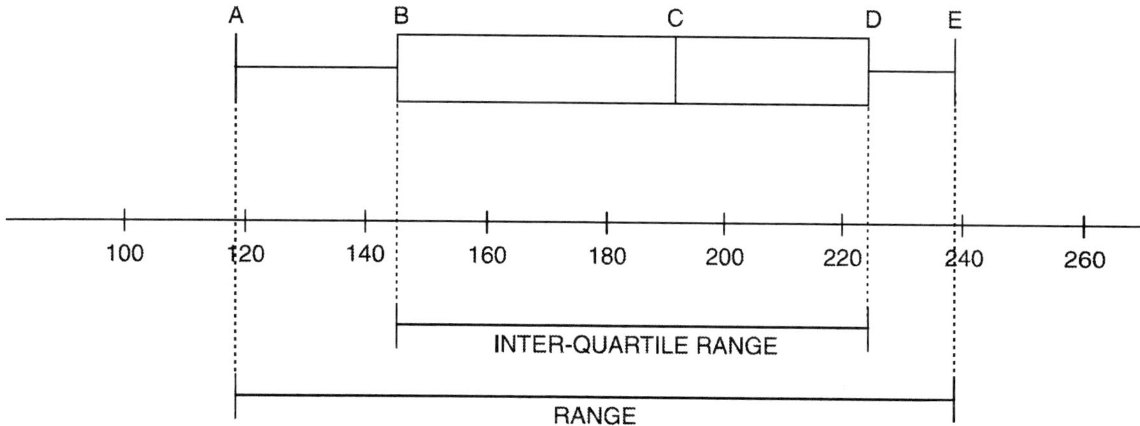

A = Lowest value in distribution (118)
B = Lower limit of inter-quartile range (145)
C = Median (192)
D = Upper limit of inter-quartile range (224)
E = Highest value in distribution (238)
A–E = Range (238 – 118 = 120)
A–B = Lower quartile (point below which 25% of distribution lie = 145)
B–D = Inter-quartile range (145 – 224)
D–E = Upper quartile (point above which 25% of distribution lie = 224)

Figure 14 Box plot

The usefulness of the box plot will become more apparent once we have looked at its features in the statistical techniques section of this chapter.

4.2 Graphical techniques – comparisons

There are a number of ways in which graphs can be used to show comparisons between individual results.

- **Contingency tables** can be used to show specific values and interdependence between values

- **Multiple bar charts** can be used to compare highest and lowest values

- **Percentage component bar charts and/or two or more pie charts** can be used to demonstrate comparison of proportions

- **Multiple line graphs** can be used to compare trends

- **Component bar charts** can be used to compare totals

- **Multiple box plots** can be used to compare distributions

We have already encountered some of these data presentation methods when we looked at the most appropriate diagrams to use if we wish to present individual results.

4.2.1 Contingency tables

Let us return to our 'holiday survey'. Suppose that we wish to determine whether there is any interdependence between the number of holidays taken and whether the respondents are male or female. We could draw up a contingency table that details our individual results as follows.

Number of holidays	Male	Female	Total
0	40	80	120
1	100	220	320
2	85	185	270
3	70	120	190
4	18	37	55
5	5	30	35
6	2	8	10
Total	320	680	1,000

Figure 15 Contingency table showing number of holidays by gender

Contingency tables are used extensively in statistical analysis. We will return to contingency tables when we study chi-square tests later in this chapter. These tests are performed in order to determine whether two or more variables are significantly associated.

4.2.2 Multiple, percentage component and component bar charts

We mentioned earlier that there were a number of different types of bar chart. We need to use these types of bar chart when comparing the following.

- Highest and lowest values
- Proportions
- Totals

If we wish to make comparisons between the highest and lowest values, the **multiple bar chart** is an ideal tool for doing this. A multiple bar chart is a bar chart in which two or more separate bars are used to present sub-divisions of data. For example, we could explore the type of holiday preferred by men and women by drawing up a multiple bar chart such as the one below.

Preferred holiday type	Male	Female	Total
Beach	40	310	350
Skiing	230	60	290
Camping	170	110	280
Other	70	10	80
	510	490	1,000

Figure 16 Contingency table showing preferred holiday type by gender

Data analysis software would use this information in order to present these results in a multiple bar chart, such as the one shown below in Figure 17.

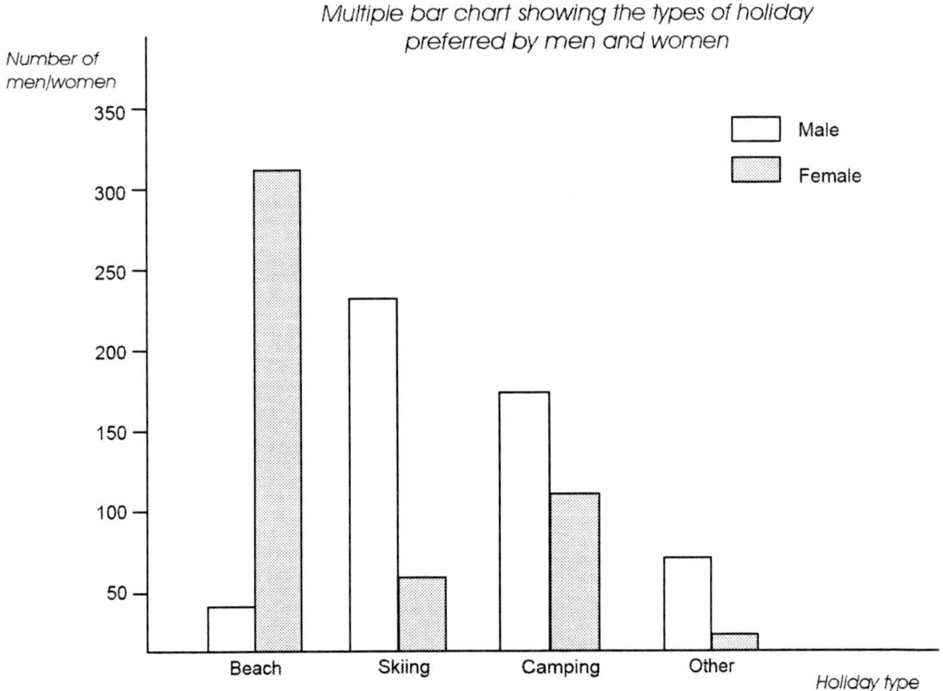

Figure 17 Multiple bar chart

Research Methods for Global Marketing Practice

If, however, we wish to make comparisons of the proportions between variables, a better tool to use would be a percentage component bar chart. A percentage component bar chart indicates the relative sizes of components by the lengths of the sections of the bar.

Suppose that we wish to show the number of holidays taken in a year by different nationalities and that the 'holiday survey' revealed the following results. We could easily make comparisons by producing a percentage component bar chart.

Number of holidays taken last year	British	American	European (excluding British)	African	Asian	Australian /New Zealand	Total
0	23	40	27	10	12	8	120
1	145	40	70	25	30	10	320
2	130	35	65	10	15	15	270
3	110	15	54	1	9	1	190
4	7	0	45	1	1	1	55
5	3	0	28	1	1	2	35
6	2	0	1	2	2	3	10
	420	130	290	50	70	40	1,000

Figure 18 Contingency table showing number of holidays taken last year and nationality

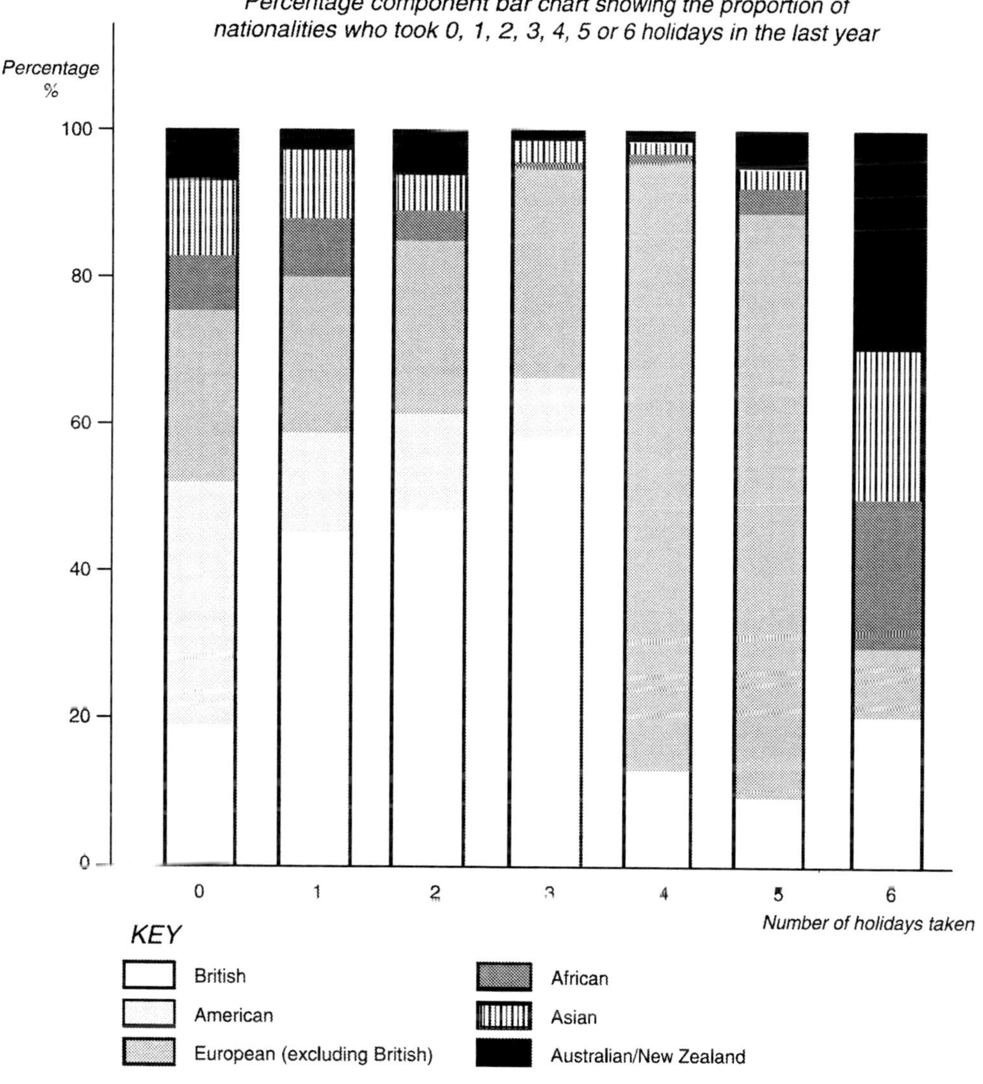

Figure 19 Percentage component bar chart

The graph above shows a lot of information and comparisons can be easily made between adjacent bars. Another way of showing this information would have been to have drawn six pie charts with one chart for 0 holidays, one for 1 holiday and so on. Within each chart, the proportion of each of the nationalities having the number of holidays in question would be shown as being coloured or shaded differently.

Component bar charts

If we wish compare totals between variables, we can use a variation on the bar chart known as the component bar chart (sometimes called a stacked bar chart). A component bar chart is a bar chart that gives a breakdown of each total into its components. The total length of each bar and each component on a bar chart indicates magnitude.

Returning to the 'holiday survey,' if we wished to show totals, the information relating to preferred holiday types of men and women that were shown in the multiple bar chart above could also be shown in a component bar chart.

The multiple bar chart indicates whether men or women take the most or least number of beach, skiing and camping holidays. On the other hand the component bar chart compares the totals between the different holiday types. It is evident from the component bar chart that beach holidays are the most popular, followed by skiing, camping and other types of holiday. In comparison, the multiple bar chart shows that beach holidays are preferred by women (by a long way!) and that skiing, camping and 'other' holidays are more popular with men.

Remember to ensure that you consciously identify what you are trying to compare before deciding the type of data presentation tool to use.

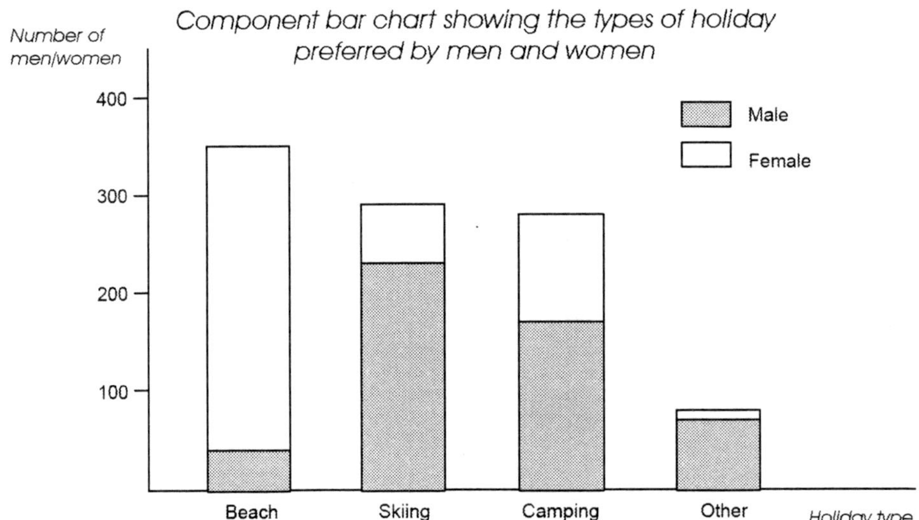

Figure 20 Component bar chart

4.2.3 Multiple line graphs

When showing the trends of individual results, we have already established that a line graph is the best way of displaying these trends. If, however, we wish to compare trends, we would simply prepare a multiple line graph. A multiple line graph is simply a graph that displays a number of trend lines on it. The point at which any of these lines cross is known as the conjunction point. We shall be looking at line graphs further when we study time series analysis in the next section of this chapter.

4.2.4 Multiple box plots

Distributions may be compared by producing multiple box plots. These are similar to the single box plot but show a number of box plots for a number of different variables. Comparisons are easily made because the individual box plots are all on the same diagram.

If we wish to determine whether a relationship exists between ranked and quantifiable data, we can plot the variables under consideration on a scattergraph.

- The horizontal or X axis of a scattergraph represents the independent variable
- The vertical or Y axis of a scattergraph represents the dependent variable.

Once these variables have been plotted, a line of best fit is drawn and its formula can be determined. The general equation for a straight line function is $Y = a + bX$.

- An upward-sloping line of best fit indicates positive correlation.
- A downward-sloping line of best fit indicates negative correlation.

Positive correlation indicates that low values of one variable are associated with low values of the other and high values of one variable are associated with high values of the other.

Negative correlation indicates that low values of one variable are associated with high values of the other, and high values of one variable with low values of the other.

Correlation can be shown graphically on scattergraphs as follows.

Perfect correlation

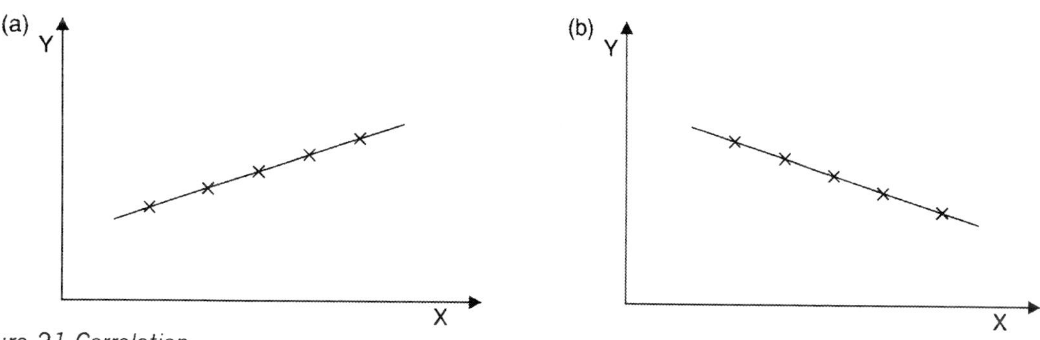

Figure 21 Correlation

All the pairs of values lie on a straight line. An exact linear relationship exists between the two variables.

Partial correlation

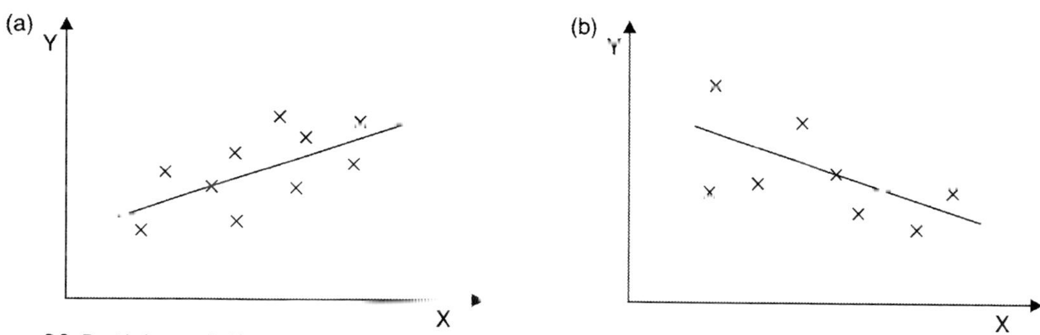

Figure 22 Partial correlation

In (a), although there is no exact relationship, low values of X tend to be associated with low values of Y, and high values of X with high values of Y.

In (b), again there is no exact relationship, but low values of X tend to be associated with high values of Y and *vice versa*.

No correlation

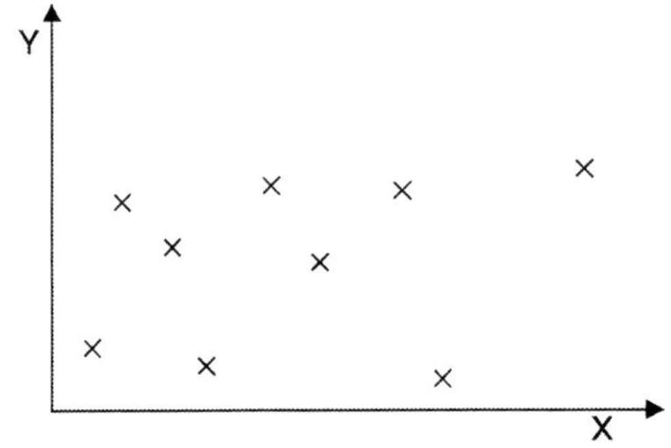

Figure 23 No correlation

We shall be looking at relationships between variables in more detail when we look at correlation and regression in section 7 of this chapter.

5 Measures of central tendency and dispersion

So far we have only considered the ways in which your research findings can be presented graphically. Statistical techniques allow you to describe and compare the data that you have collected.

Statistical techniques fall into the following groups.

- Measures of central tendency
- Measures of dispersion
- Existence of relationships
- Existence of trends

5.1 Measures of central tendency

Measures of central tendency are more commonly known as averages. Averages provide a general impression of the most common values and are representative of the population or sample. There are three main types of average.

- Mean
- Mode
- Median

5.1.1 Arithmetic mean

The arithmetic mean is the only average that is calculated by taking all of the data items in a distribution into account. It is often written as $\bar{x}$. The arithmetic mean has several benefits.

- Simplicity of calculation
- Universally understood
- Represents the entire data set
- Use in statistical analysis

5.1.2 Mode

The mode or modal value is an average representing the most frequently-occurring value or category. Unlike the arithmetic mean and the median, it can be used for both categorical and quantifiable data.

For example, in relation to the 'holiday survey', the preferred accommodation of those respondents who holiday in Britain is self-catering (refer to the simple bar chart in Figure 9 above) and therefore the modal holiday accommodation is self-catering.

The mode affords various benefits.

- Easy to determine
- Not influenced by any extreme values in the data set
- Can be used for categorical (ordinal) data (unlike the mean and the median)
- Can represent an actual item in the distribution

5.1.3 Median

The median or middle value is found by arranging the data items in your distribution in order of magnitude. (A list of data items in order of value is known as an array). The item in the middle of the range is known as the median.

If there are an odd number of items in the array, it will be easy to find the middle item. If, however, there are an even number of items in the array, there will be two middle items and you will need to calculate the arithmetic mean of these two items.

Benefits of the median are as follows.

- Easy to understand
- Not affected by extreme values
- Can represent an actual item in the distribution

5.2 Measures of dispersion

In addition to describing a variable's average value it is also useful to describe how dispersed around the average the variables are. The main measures of dispersion are as follows.

- Range
- Inter-quartile range
- Quartiles
- Deciles and percentiles
- Spread
- Standard deviation
- Coefficient of variation

5.2.1 Range

The range is the difference between the highest and the lowest values in a distribution. For example, in our 'holiday survey', the number of holidays taken by respondents ranged between 0 and 6 (see Figure 15). The range of these variables is therefore 6 (6-0). While this statistic is not frequently used in research reports, it does give the researcher a good idea of the diversity of the values that he is dealing with.

5.2.2 Inter-quartile range

We had a brief look at the inter-quartile range and the median when we studied box plots in the previous section of this chapter. The following diagram shows how the range can be divided into a number of different parts.

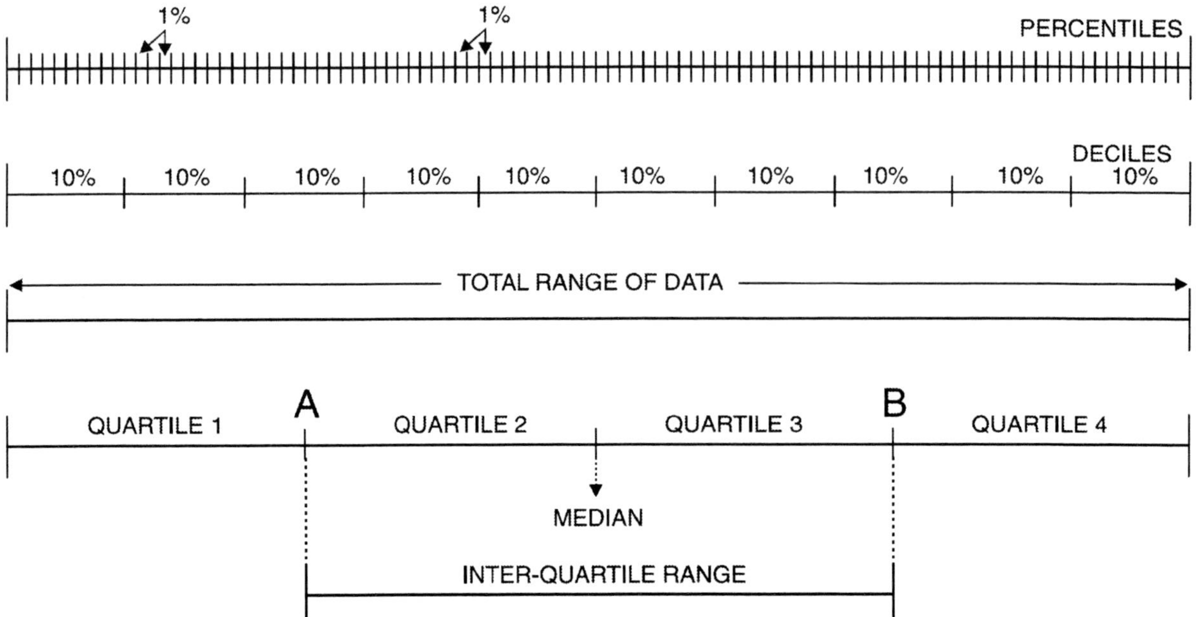

Figure 24 Range, percentiles, deciles and quartiles

The lower quartile is represented by point A in the graph above and the upper quartile is represented by point B.

- The lower quartile is the point below which 25% of the population lie
- The upper quartile is the point above which 25% of the population lie

The inter-quartile range is a frequently used statistic which identifies the middle 50% of the population. The median is the middle point in the inter-quartile range.

It is also possible to divide the range into deciles and percentiles.

- Percentiles divide a distribution into 100 equal parts
- Deciles divide a distribution into 10 equal parts

5.2.3 Standard deviation

An important part of your research will involve identifying the extent to which data values for a variable are spread around the mean. Data values which are all close to the mean indicate that the mean is a typical value of all of the data items in the distribution.

The standard deviation is a statistical value that describes the extent to which data items in a population are spread around the mean. The standard deviation of data items from a sample (rather than a population) is calculated using a slightly different formula. If, however, your sample contains more than thirty data items, the statistics calculated are more or less the same.

The standard deviation is calculated by taking the square root of the variance. The variance is the average of the squared mean deviation for each value in a distribution.

If you are using computerised analysis software, it will automatically calculate the standard deviation of the data that you have collected, so you don't need to worry about any complicated formulae here! You must, however, be able to interpret the statistics that the software calculates for you.

5.2.4 Coefficient of variation

If you wish to compare the spread of data items around the mean for different distributions, you will need to calculate a statistic known as the coefficient of variation. The higher the value of the coefficient of variation, the more spread out the data in the distribution. The coefficient of variation is calculated by dividing the standard deviation of the distribution by the arithmetic mean of the distribution.

The coefficient of variation is also known as the coefficient of relative dispersion.

Suppose you collect two sets of data, A and B. The statistical software programme that you are using might provide you with the following results.

	Data set	
	A	B
Mean	120	125
Standard deviation	50	51
Coefficient of variation	0.417	0.408

Figure 25 Coefficient of variation table

5.2.5 Interpretation of results

- Data set B has the higher standard deviation in absolute terms (51 compared to 50)

- Data set A has the higher coefficient of variation indicating that A has a greater relative dispersion than data set B.

It can therefore be concluded that data set A is more spread out than data set B (because it has a higher coefficient of variation).

6 Hypothesis testing

One of the most important uses of statistics is to investigate whether relationships exist between one variable and another. There are a number of ways in which statistics can be used in order to examine this. Hypothesis testing is one of these ways.

Selecting which statistical test to choose is often one of the major areas of concern when completing a student research project. Complex-looking statistical formula fill these sections of text books and can be highly daunting if you are new to the area of statistics. In reality, so long as a logical procedure is followed, this is one area which can be easily covered and should not be feared. The use of computers has meant that even if you are not a mathematician, you can get through your research project with a basic working knowledge.

The following steps should be taken.

Step 1   Specify the null hypothesis

Step 2   Specify the alternative hypothesis

Step 3   Establish the level of statistical significance

Step 4   Select the appropriate statistical test

Step 5   Test for statistical significance

Steps 1 and 2 – Specify the null and alternative hypotheses

Definition

The **null hypothesis** stipulates that the two variables are not related. This is the hypothesis which is tested.

The **alternative hypothesis** states that there is a significant association between the variables and the results obtained did not occur by chance.

Example null and alternative hypotheses used within the airline industy may be:

Null (Ho) There is no difference in levels of customer service between ground and cabin crew.

Alternative (HA) The levels of customer service differs between ground and cabin crew.

Step 3 – Establish the level of statistical significance

Statistical significance enables the researcher to assess whether in fact the findings are generalisable to the population. If a sampling error has occurred (even when probability sampling has been used correctly) the sample will be unrepresentative of the wider population and findings will be invalid (Bryman and Bell, 2007). What you as a researcher can do however is to provide an indication of how confident you can be in your findings.

Consider medical research for a moment. In conducting tests on a new drug we would hope that researchers would be highly confident from their findings that the drug is not harmful to humans. If we think about how confident we would like them to be in percentage terms we probably would say we expect 100% confidence but realistically would settle for 99% confidence.

In the research that you are conducting, although it is highly important to your organisation (and indeed you in enabling you to pass your qualification!), within business-related subjects it is normal to accept that the researcher is 95% confident in their findings.

So what does all this mean in statistical terms?

Definition

Statistical significance refers to '*the level of risk you are prepared to take that you are inferring that a relationship exists between two variables in the population from which the sample was taken when in fact no such relationship exists*' (Bryman and Bell, 2007).

By significance we are not implying something is *important* but how *confident* we are in the findings.

In business research the convention is to say that there are up to 5 chances in 100 that we might reject the null hypothesis (therefore saying that a relationship does exist) when it should have been confirmed. This type of error is known as a Type I error.

Type II errors also exist and refer to us failing to reject the null hypothesis when it should have been rejected.

The significance level is denoted by p which stands for probability (hence the link with sampling theory).

If we accept a significance level of 95% confidence with a 5% chance of wrongly rejecting the null hypothesis this is shown as $p<0.5$.

The following table adapted from Bryman and Bell (2007) shows the relationship between significance level and likelihood of making the two error types.

Research Methods for Global Marketing Practice

Significance level	Type I error	Type II error
	Rejecting null hypothesis when it is true	Confirming the null hypothesis when it should be rejected
$p<0.5$ (95% confident)	Greater risk	Lower risk
$p<0.1$ (99% confident)	Lower risk	Greater risk

Consider the following scenario when measuring perceptions of a new film.

Null hypothesis – there is no difference between male and female cinema goers' perceptions of the film.

Statistical tests show that there is no significant difference and so we reject this hypothesis and accept the alternative hypothesis that there is a difference between male and female cinema goers' perception of the film.

At 95% confidence where there is a 5% chance ($p<0.5$) that actually the null hypothesis is correct and there wasn't a difference in perceptions after all.

Once you have identified your level of confidence you can select an appropriate test for statistical significance.

Step 4 – Select the appropriate statistical test

In choosing the test, there are three important considerations as shown below.

Consideration 1 – How many groups are we comparing?

If we are testing the significance level of just one variable, eg all visitors to a cinema then we would conduct univariate analysis tests.

If we wanted to compare between males and females then we would have two groups and use bivariate tests.

If we were comparing between infrequent cinema goers, monthly visitors and weekly visitors then we would have three groups and use multivariate analysis.

Consideration 2 – Are the groups related or independent?

Sometimes groups have no effect on the other but at other times one group can influence the results of the second group.

Independent samples – the measurement of the variable of interest in no way affects the measurement of the variable in the other sample. For example, in a survey among cinema goers, a comparison is made between male and female perceptions of the film (where there has been no discussion between respondents so no influence can be exerted by either gender).

Related samples mean that the measurement of the variable of interest in one sample can influence the measurement of the variable in the other sample. For example, if we were measuring responses to a taste test for cheese the residual taste of the first cheese may influence the taste of the second cheese.

Consideration 3 – What type of data do I have?

At the start of the chapter we began by looking at the alternative types of data and their associated mathematical properties (nominal, ordinal, interval and ratio). Identify which level of measurement is used.

Now you have considered the nature of the data, you can use the following three tables to select the appropriate test.

Table A – Statistical tests for single group

Nominal	Ordinal	Interval / ratio
Chi-square test Binominal test	Kolmogorov-Smirnov one sample test	z-test (samples size less than 30) t-test (samples larger than 30)

Table B – Statistical tests for two groups

Sample related or independent?	Nominal	Ordinal	Interval / ratio
Related	McNemar test	Wilcoxon matched pairs sign test	t-test for related groups
Independent	Chi-square Z-test for dichotomous variables	Mann-Whitney median Kolmogorov-Smirnov two sample test	z-test (samples size less than 30) t-test (samples larger than 30)

Table C – Statistical tests for more than two groups

Sample related or independent?	Nominal	Ordinal	Interval / ratio
Related	Cochran Q test	Friedman test Kendall coefficient of concordance	Repeated measures analysis of variance
Independent	Chi-square	Kruskall-Wallis test	Analysis of variance

Tables based on diagrams in Dillon *et al* (1994).

Activity 2

Using the tables above identify the appropriate statistical test to use within the following scenario.

The views of brand managers for an international drinks manufacture have been gathered via an employee survey about their view of the new global advertisign campaign.

The question posed which need to be analysed was:

Please score out of 10 the likelyhood of the campaign to achieve our communications objectives. ____/10

You would like to compare the views of brand managers in Europe, Africa, Asia and the US.

Step 5 – Test for statistical significance

It is at this point that you are ready to complete the actually statistical analysis.

 Research Methods for Global Marketing Practice

> ## Assessment advice
>
> Although we also include mathematical formula within this Study Text, it is there to demonstrate to you the mathematical thinking behind your analysis. It is highly unlikely that you will be completing these calculations yourself because via the Anglia Ruskin e-Vision system you will be able to download the data analysis software SPSS. Please do not spend unneccessary hours working though the actual calculations but attempt to gain an overall understanding of which statistical tests you would use in which circumstances so that you can put together a comprehensive analysis plan.
>
> There will be additional Fact Sheets and other resources available on the VLE when you are at the stage of analysing your data.

We shall now look at some of the tests you may use (again this is most likely to be completed in SPSS rather than by hand).

6.1 Chi-square test

6.1.1 Distinction between univariate and multivariate tests

The chi-square test can be used on univariate or multivariate frequency distributions. As previously explained, univariate data analysis is where you are only examining one variable. Multivariate data analysis is where you are looking at two or more variables. The study of just two variables is known as bivariate data analysis.

6.1.2 Univariate test-scope for use

As discussed earlier, you might want to present some of the data you have analysed in the form of a frequency distribution. Having done that you would like some indication as to whether 'the difference between the observed frequency distribution and the expected frequency distribution can be attributed to sampling variation'. (Zikmund, 2002)

According to Diamantopoulos and Schlegalmilch (2000) when you compare a set of observed frequencies with a set of theoretical frequencies, the chi-square test is the one to use. The question then arises whether the differences between observed and theoretical frequencies are significant.

Phelan and Reynolds (1996) explain that the chi-square test is often referred to as a test of 'goodness of fit,' a description which you are likely to encounter in textbooks on quantitative techniques for business research. They cite Hammerton (1975), in describing the technique as 'one of the most widely used of all the statistical techniques available to the behavioural scientist'.

Phelan and Reynolds, *ibid*, classify chi-square tests into two categories.

- Those used on frequency distributions (univariate scenario).
- Those used on contingency tables (multivariate scenarios).

Collis and Hussey (2003) explain the objectives of a chi-square test quite briefly and clearly stating that it allows you to draw statistically valid inferences as to whether there are any significant differences between the actual (observed) frequencies and hypothesised (expected) frequencies. In other words, whether the differences are due to some underlying, universal difference or merely to chance.

6.1.3 Process involved

Here is a process drawn from various sources – Weiss (2002) and Diamantopoulos and Schlegalmilch (2000).

Step 1   Set up the null and alternative hypothesis; as above.

Step 2   Decide on significance level of the test. This is usually 5%.

Step 3 Calculate the chi-square statistic for the sample.

Where   O = Observed frequencies
            E = Expected frequencies

and all the cells in the table are added up.

Step 4 Determine critical value beyond which null hypothesis is rejected and alternative hypothesis is adopted.

This is done using chi-square distribution tables. Here is a short extract.

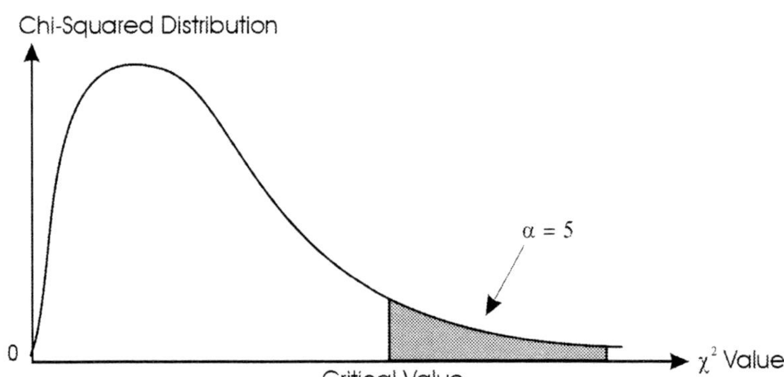

Degrees of Freedom (d.f.)	Area in Shaded Right Tail (α)		
	.10	0.5	0.1
1	2.706	3.841	6.635
2	4.605	5.991	9.210
3	6.251	7.815	11.345
4	7.779	9.488	13.277
5	9.236	11.070	15.086
6	10.645	12.592	16.812
7	12.017	14.067	18.475
8	13.362	15.507	20.090
9	14.684	16.919	21.666
10	15.987	18.307	23.209

Figure 26 Chi-square test

- Degrees of freedom (d.f.) refers to the number of values/observations that may be varied in the final calculation of a statistic without compromising the assumptions on which it is based. You do not need to actually understand the mathematics behind this concept; you just need to know how to calculate it.

  For a one sample/one way test:

  d.f.   = (r – 1) (c – 1)

  where r = the number of rows in the table (excluding totals)
  and c   = the number of columns in the table (excluding totals)

  Where you are faced with a univariate frequency table, the (c – 1) part of the formula drops out. It does **NOT** become 1 – 1 = 0, which would cause v to also be 0.

  Here is another way of trying to understand the idea of degrees of freedom. Consider adding five numbers together where the total value is known. Eg

$$\begin{array}{r} 3 \\ 7 \\ 1 \\ 4 \\ \underline{X} \\ \underline{20} \end{array}$$

The fifth number, x, must be 5 but the first four numbers may be varied. Hence there are five numbers but four degrees of freedom involved in this example.

- Read off **'critical value'** from tables.

Step 5   Evaluate hypothesis; compare critical value per tables with chi-square value per calculations

- Computed chi-square value > Critical value per tables → Reject null hypothesis
- Computed chi-square value ≤ Critical value per tables → Accept null hypothesis

Step 6   Express conclusion in words

Your research survey includes questions that establish the profile of spectacle wearers and non-spectacle wearers in City A. You analyse your data arising from your sample and produce the following frequency distribution.

	Observed frequencies
Spectacle wearers	1,440
Non-spectacle wearers	1,560
TOTAL	3,000

You would like to know whether the difference between the above distribution and the expected distribution is due to some underlying reason or merely the result of chance.

Step 1   Set up hypotheses

Null hypothesis: There is no difference between the observed frequencies and the expected frequencies.

Alternative hypothesis: The difference between observed and expected frequencies is statistically significant due to matters other than chance

Step 2   Significance level of the test

The test will be performed at the 5% level of significance

Step 3   Application of formula

Eyesight status	Observed frequency (O)	Expected frequency (E)	(O – E)	$\dfrac{(O-E^2)}{E}$
Spectacle wearer	1,445	1,500	– 55	$\dfrac{3,025}{1,500} = 2.017$
Non-spectacle wearer	1,555	1,500	+ 55	$\dfrac{3,025}{1,500} = 2.017$
	$\underline{3,000}$	$\underline{3,000}$		X2 = $\underline{4.034}$

Step 4   Determination of critical value

d.f. = 2 – 1 = 1

Critical value @ 5% and 1 d.f. = 3.841

Step 5	Evaluation of hypothesis

Step 5   Evaluation of hypothesis

4.034 > 3.841 $\therefore$ Reject null hypothesis.

The alternative hypothesis applies and the variables are associated with each other.

Step 6   Expression of conclusion

The differences in frequency between spectacle wearers and non-spectacle wearers is statistically unlikely to be due to chance, but probably due to some underlying factor or factors.

The difference is significant at the 5% level.

6.1.4 Multivariate scenario – scope for use

When analysing the data you have collected, you may produce some contingency tables. You might wish to draw some statistical inferences on the comparison of observed frequencies with expected frequencies.

6.1.5 Hypothesis involved

Null hypothesis: There is no difference between the frequencies observed and expected relative frequencies.

Alternative hypothesis: The difference between the samples are statistically different.

Diamantopoulos and Schlegelmilch (2000) explain that the reason for focusing on relative rather than absolute frequencies is that the group may have unequal sample sizes which must be taken into consideration.

Global case study

Your survey on eyesight reveals the following bivariate analysis. The contingency table shows a comparison between gender and lens type.

Type of lens	Male	Female	Total
Single vision	360	400	760
Bifocal	200	225	425
Varifocal	140	115	255
Total	700	740	1,440

You might also have produced a percentage component bar chart to support your contingency table.

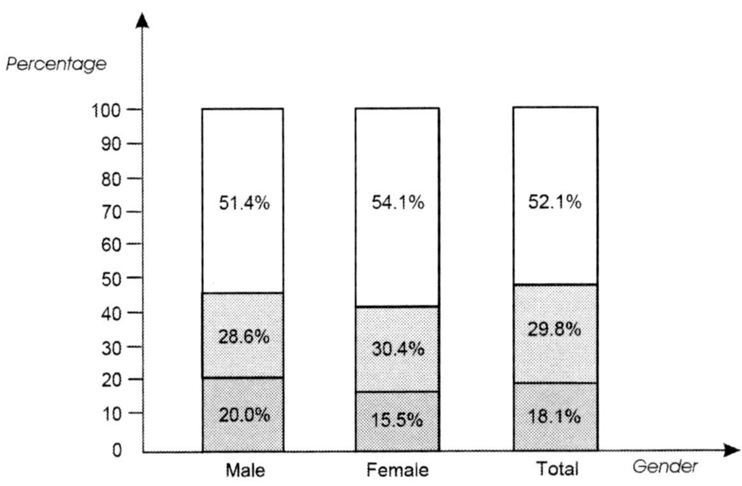

Research Methods for Global Marketing Practice

You would like to know if the difference between male and female usership patterns is statistically significant.

Based on the relative frequencies for the total column you are able to develop a table of expected frequencies for each gender category.

Type of lens	Male	Female	Total
Single vision	364.7	385.5	750
Bifocal	208.6	220.5	430
Varifocal	126.7	133.9	260
Total	700	740	1,440

Step 1 **Hypothesis involved**

Null hypothesis: There is no difference between the levels of male and female usership over the various types of lenses as compared to the expected frequencies.

Alternative hypothesis: There is a statistically significant difference between male and female usership over the various types of lenses in relation to expected frequencies.

Step 2 **Significance level for the test**

The test will be performed at the 5% level of significance.

Step 3 *Application of formula*

Cell	Observed frequency (O)	Expected frequency (E)	(O – E)	$\dfrac{(O-E^2)}{E}$
1	360	364.7	(4.7)	0.061
2	400	385.5	14.4	0.726
3	200	208.6	(8.6)	0.355
4	225	220.5	4.5	0.092
5	140	126.7	13.3	1.396
6	115	133.9	(18.9)	2.678
	1,440	1,440.0		$X^2 =$ 5.308

Step 4 **Determination of critical value per table**

$$
\begin{aligned}
\text{d.f.} \;&=\; (r-1) \quad (c-1) \\
&=\; (3-1) \quad (2-1) \\
&=\; 2
\end{aligned}
$$

Value per tables = 5.991

Step 5 **Evaluation of hypothesis**

5.308 < 5.991

ie Computed statistic for samples < Critical value per tables

Hence, null hypothesis is sustained.

Step 6 **Expression of conclusion**

At a 5% level of significance, there is no difference between the levels of male versus female usership over the various types of lenses in relation to expected usership.

Here is an example using the software-driven approach.

Looking back at our 'holiday survey', we might wish to determine whether there is any association between preferred holiday types and gender. Nominal data (such as the data shown in Figure 27 below) is summarised in a two-way contingency table.

In order to establish whether the two variables are independent or associated with each other, you simply need to choose the chi-square statistic and then run the software program.

To test whether there is any significant association between preferred holiday type and gender we can use the data in Figure 27 as shown below.

Preferred holiday type	Male	Female	Total
Beach	40	310	350
Skiing	230	60	290
Camping	170	110	280
Other	70	10	80
	510	490	1,000

Figure 27 Contingency table – preferred holiday type and gender

The statistical analysis software will use the data in Figure 27 in order to create a new data matrix which gives the information as shown in Figure 28 below.

Preferred holiday type	Male	Female	Column total	Expected %
Beach	40	310	350	35%
Skiing	230	60	290	29%
Camping	170	110	280	28%
Other	70	10	80	8%
Row total	510	490	1,000	100
Expected %	51%	49%		

CHI-SQUARE = 365.54
DEGREES OF FREEDOM = 3
PROBABILITY = less than 0.001*

Figure 28 Chi-square test – significant association between preferred holiday type and gender

The table above shows that there is an overall chi-square value of 365.54 with three degrees of freedom. This is greater than the value from the table at 5% and 3 degrees of freedom → 7.815. The conclusion is that the differences between the variables is statistically significant and not owing to chance.

6.2 t-tests

6.2.1 Scope for use

t-tests are a handy tool for helping you to draw statistical inferences about the mean scores you may have obtained from analysing your quantitative data.

6.2.2 Criteria for use

When analysing the quantitative data you have collected, there may be situations where:

- The sample size ($n \leq 30$) is not large enough to allow the population mean to be approximated by a normal distribution.

Research Methods for Global Marketing Practice

- The population standard deviation is unknown.

In such circumstances the so-called z-test is used.

6.2.3 Univariate or bivariate

As with the chi-square testing, t-tests can be used on univariate or bivariate data, ie in practice, applied on a single variable frequency distribution or on a bivariate contingency table.

6.2.4 Hypothesis involved

In a **univariate t-test** you would be testing whether the difference between the sample mean arising from your research and the population mean is statistically significant. Again, in practice, you would be seeking to express your results statistically at a 5% significance level.

As before, the process involves setting up a null hypothesis and an alternative hypothesis.

In a **bivariate t-test** you would be interested in whether the difference between the mean scores arising from two independent samples is statistically significant. Again, this test is done at a 5% significance level. And again, you would proceed by setting up a null hypothesis and an alternative hypothesis.

6.2.5 Univariate scenario

As discussed previously, this scenario refers to research looking at one variable, eg MSc salaries.

Global case study

You believe that MSc GMP qualified marketers earn an average remuneration package of $100,000. You conduct a random sample of 25 GMP graduates, asking them to provide you details of their remuneration.

From the data you have collected, you calculate a sample mean, $\bar{x}$ = $110,000 and a sample standard deviation of $5,500

Step 1   **Set up a null and alternative hypothesis**

Null hypothesis:	The population mean μ = $100,000
Alternative hypothesis:	The population mean $\mu \neq$ $100,000

Step 2   **Significance level for the test**

This is usually set at 5%

Step 3   **Calculation of the t-statistic for the sample**

$$t = \frac{\bar{x} - \mu}{s_{\bar{x}}}$$

where
$$\bar{x} = \text{Sample mean}$$
$$\mu = \text{Population mean}$$
$$s_{\bar{x}} = \frac{s}{\sqrt{n}} \quad (s_{\bar{x}} \text{ is called the Standard Error of the Mean})$$
$$s = \text{Sample standard deviation}$$
$$n = \text{Sample size}$$

$$\text{Hence } t = \frac{\$110,000 - \$100,000}{\$1,100}$$

$$= \frac{\$10,000}{\$1,100}$$

$$= \underline{\underline{9.09}}$$

Step 4   **Identification of critical value from t-distribution tables**

$$\text{Degrees of freedom (d.f.)} = n - 1$$
$$= 25 - 1$$
$$= \underline{\underline{24}}$$

Read off critical value from tables at 5% significance level and 24 degrees of freedom.

$$\text{Critical value} = \underline{\underline{1.711}}$$

Step 5   **Evaluation of hypothesis**

t-value of sample > Critical value per t-distribution

Hence null hypothesis is not sustained. Alternative hypothesis applies.

Step 6   **Expression of conclusion**

At a 5% significance level, the average remuneration package for GMP graduates is not equal to $100,000

6.2.6 Bivariate scenario

Often you will be dealing with more than one variable and wish to draw some statistically supported inferences about mean values shown in a contingency table. For example, instead of just looking at average remuneration packages, you might wish to compare the average rewards of male versus female GMP graduates.

The t-test technique is sometimes also used to compare the results arising from *experimental groups* against the results of *control groups*. However, there are often practical issues involved in comparing experimental groups with control groups resulting from underlying differences such as age, experience, skills, attitudes and so on, of group members. You should consult properly with your E-Tutor if you wish to pursue a *matched pairs technique.* In this module, we shall assume you have used independent or unrelated samples.

6.2.7 Hypotheses involved

As before, you would set up two hypotheses.

Null hypothesis: There is no difference between the mean values for the two groups

Alternative hypothesis: The difference between the means for the two groups is not due to chance, but is statistically significant.

6.2.8 Basic assumptions

Zikmund (2002) specifies two assumptions for applying the t-test for difference of means.

- The two samples are drawn from Normal distributions.

- Because the standard deviations of two populations are unknown, it is assumed that the variances of the two populations or groups is equal.

The computations associated with t-tests for difference of means are somewhat complicated and it is recommended that you use appropriate software, in consultation with your supervisor.

Remember the original null hypothesis is that there should be no difference between the population means. Hence there should be no difference between the sample means. The proposition is expressed in the following equation.

$$t = \frac{\text{Mean 1} - \text{Mean 2}}{\text{Variability of Random Means}}$$

$$= \frac{\bar{X}_1 - \bar{X}_2}{S\bar{X}_1 - \bar{X}_2}$$

A pooled estimate of the standard error is a better estimate of the standard error than one based on the variance from either sample. This required the assumption that variances of both groups (populations) are equal.

(Zikmund, 2002)

Here is the formula to calculate the pooled standard error of the difference between means of independent samples:

$$S\bar{X}_1 - \bar{X}_2 = \sqrt{\left(\frac{(n_1 - 1)S_1^2 + (n_2 - 1)S_2^2}{n_1 + n_2 - 2}\right)\left(\frac{1}{n_2} + \frac{1}{n_2}\right)}$$

where S_1^2 = the variance of Group 1

S_2^2 = the variance of Group 2

n_1 = the sample size of Group 1

n_2 = the sample size of Group 2

Global case study

You have collected information on the remuneration packages of marketers who also hold an MSc and marketers not holding an MSc. (Data has been collected to the nearest $1,000.)

You have taken a simple sample from two populations and produced the following analysis

Marketers with MSc		Marketers without MSc	
$\bar{X}_1$	= $160	$\bar{X}_2$	= $140
$\bar{S}_1$	= 23	S_2	= 27
n_1	= 25	n_2	= 15

Step 1   **Set up a null and alternative hypothesis**

Null hypothesis: There is no difference between the mean remuneration packages of marketers with MSc versus marketers without an MSc.

Alternative hypothesis: The difference between the mean remuneration packages of marketers with MSc versus marketers without MSc is statistically significant and not due to chance but reflects underlying factors

Step 2   **Decide on a significance level for the test**

This is usually set at 5%

Step 3   **Apply the formula**

$$S\bar{x}_1 - \bar{x}_2 = \sqrt{\left[\frac{(25-1)23^2 + (15-1)27^2}{25+15-2}\right]\left[\frac{1}{25}+\frac{1}{15}\right]}$$

$$= \sqrt{\left[\frac{(24\times23^2)+(14-27^2)}{38}\right]\left[\frac{3+5}{75}\right]}$$

$$= \sqrt{\frac{12{,}696+10{,}206}{38}\times\frac{8}{75}}$$

$$= \sqrt{64.286}$$

$$= 8.018$$

$$\text{Hence } t = \frac{160-140}{8.018}$$

$$= 2.49$$

Step 4   **Determine critical value per t-distribution tables**

Test d.f. $= 38$

Per tables at 5% significance:

35 d.f.	$= 1.690$
40 d.f.	$= \underline{1.684}$
	$\underline{(0.006)}$

$$\text{Hence } t\text{-value per tables} = 1.690 + \left(\frac{0.006}{5}\right)2$$

$$= 1.692$$

Step 5   **Evaluation of hypothesis**

t-value of samples > Critical value per t-distribution

Hence null hypothesis is not sustained. The alternative hypothesis applies.

Step 6   **Expression of conclusion**

At 5% significance level, the difference between the mean remuneration of marketers with an MSc and marketers without an MSc are statistically significant.

Global case study

Here is an example making use of software.

Suppose that we wish to investigate whether there is a **significant difference** in the average time taken (in minutes) by a sample of men and women to look up a particular holiday company on the internet based on the following results.

	Men	Women
Number in sample	6	8
Mean time (in minutes)	4.26	4.35
Standard deviation	0.0548	0.0424

t-value	$= 3.46$
Degrees of freedom	$= 12$
Probability	$=$ Less than 1%

Figure 29 t-test results

Research Methods for Global Marketing Practice

Your software will produce results shown above and it is then up to you to interpret them (using the t-distribution tables). We can interpret the results shown above as follows.

As a rule of thumb, '*if two variables are significantly different, this will be represented by a larger t-statistic with a probability less than 0.05*' (Saunders *et al* (2009)). From t-distribution tables, where we have 12 degrees of freedom and a probability of 0.05, the t-value is 2.18. In this example, we have a t-value of 3.46 which corresponds to a probability of less than 0.05 (it is actually somewhere between 0.005 and 0.0005). Such results indicate that the average times taken by men and women to look up the tour company on the internet were significantly different.

There are a couple of assumptions associated with t-tests that you should be aware of.

* They assume that the data is Normally distributed.
* It is assumed that the two groups have the same (or very similar) variances.

Saunders *et al* (2009) explain that:

'*Although the t-test assumes that the data are Normally distributed this can be ignored without too many problems even with sample sizes of less than 30. The assumption that data for the two groups have the same variance (standard deviation squared) can also be ignored provided that the two samples are of similar size.*'

6.3 Analysis of variance

6.3.1 Scope for use

During your research, you might have collected data on several independent random samples. You may wish to draw inferences whether the means relating to three or more groups of data are significantly different. In these circumstances because there are three groups, a t-test cannot be employed to determine statistical significance. You can test whether these groups are significantly different by using a one-way ANOVA test (one way analysis of variance).

Global case study

You have collected data on the remuneration packages of the following groups.

* Managers with only one qualification
* Managers with MSc
* Managers with MSc, plus doctorate

You wish to draw some statistically valid inferences about their average earnings.

6.3.2 Underlying approach

ANOVA calculates an F-ratio or F-statistic which analyses:

* Variations arising within groups
* Variations arising between groups

by comparing means.

6.3.3 Hypotheses involved

Null hypothesis: There is no difference between the means; they are mutually equal.

Alternative hypothesis: The difference between the means for the various groups is not due to chance, but is statistically significant.

6.3.4 Assumptions involved

Kazmievr (2003) sets out two underlying assumptions for using the analysis of variance technique.

- The various underlying populations are Normally distributed.
- The various underlying populations have the same variance.

6.3.5 Statistical process

Referring back to the results of our 'holiday survey' we might wish to establish whether there are significant differences in the different nationalities (categorical data) and annual incomes (quantifiable data) of respondents.

The ANOVA would therefore calculate the mean annual salary for each nationality and analyse any variations within each category (nationality). Then, the test would go one step further and analyse any variations between the different categories. The result, which is given in the form of an F-statistic and an F-probability must be interpreted.

Because of the complexity of the mathematics involved, we would recommend that you use appropriate software to conduct your tests. Do this in consultation with your supervisor.

Figure 30 below shows output from statistical analysis software for an ANOVA test.

Source	Degrees of freedom	Sum of squares	Mean squares	F-statistic	F-prob
Between groups	X	X	X	F	<0.001
Within groups	X	X	X		
Total	X̄	X̄			

Figure 30 ANOVA output

6.3.6 Evaluating results

In general, there are indications that the groups, being tested are significantly different if:

- F-statistic is a high value.
- F-probability is less than 0.05 (5%).

---

7 Existence of relationships

We will now turn our attention to the statistical tests that you can use in order to establish whether relationships exist between two variables. The area of statistics that is concerned with this is known as correlation and regression.

7.1 Correlation and regression

Measures of correlation and regression can be thought of as the two-variable equivalents of the one-variable measures of centrality and dispersion.

- **Regression** is a technique for identifying the linear association between a dependent variable and an independent variable. It locates two-variable data in terms of a mathematical relationship where the general form is $y = a + bx$ which can be depicted as a graph. Here y is the dependent variable and x is the independent variable. A and b are constants within the equation (Zikmund, 2002). The primary objective is to estimate the value of the dependent variable given the value of the independent variable.

- **Correlation**, on the other hand, measures the degree of relationship between variables. (Kazmier 2003)

For the purposes of this module, we will restrict our coverage to scenarios involving only one independent variable and one dependent variable. A linear model will be assumed for regression and correlation analysis.

7.1.1 Degrees of correlation

When the value of one variable is related to the value of another, they are said to be correlated. There are varying degrees of correlation.

- Perfect positive correlation
- Perfect negative correlation
- Partial positive correlation
- Partial negative correlation
- No correlation

At this stage it is probably opportune to visit Walkup's Law No. 3 of statistics which states:

'Unless you can think of a logical reason why two variables should be connected as cause and effect, it doesn't help much to find a correlation between them. In Columbus, Ohio, the mean monthly rainfall correlates very nicely with the number of letters in the names of the months (to which they relate)!'
(cited in Zikmund, 2002)

If you wish to determine whether a relationship exists between two variables, you will need to run a program that will calculate a statistic known as a **correlation coefficient**. The value of this statistic will enable you to determine the strength of any relationship which might be in existence.

7.1.2 Evaluating results

The correlation coefficient (denoted by the symbol r) can take on any value between -1 and $+1$. The following diagram shows how to interpret the strength of a relationship between two variables once you have calculated a **correlation coefficient**.

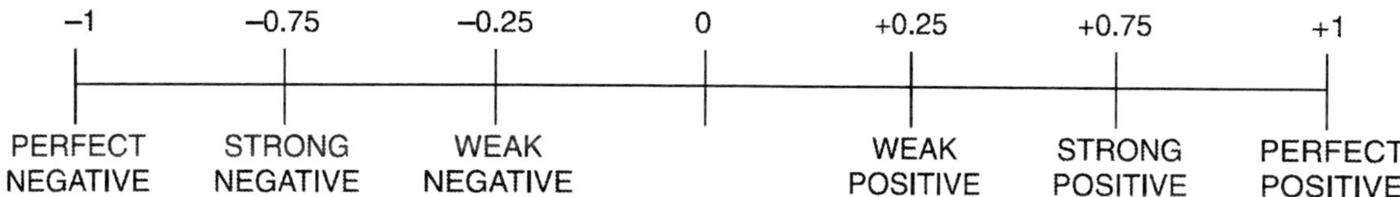

Figure 31 Interpreting correlation coefficient values

A correlation coefficient of 0 indicates that the variables in question are **totally independent** of each other and that **no relationship** exists.

The warning of Zikmund (2002) should be borne in mind that correlation does not prove causation, as variables other than those being measured may be involved.

Here are some examples of variables which might be correlated.

- A person's height and weight
- Number of customer calls and sales volume
- Fertiliser use and crop yield
- Atmospheric temperature and beer sales

7.2 Pearson's product moment correlation coefficient

7.2.1 Scope for use

Pearson's product moment correlation coefficient (PPMCC) helps you to establish whether a relationship exists between two variables and if so, how strong the relationship is.

7.2.2 Assumptions involved

Collis and Hussey (2003) outline some of the conditions governing the application of PPMCC.

- The data must be interval or ratio status and nominally distributed
- The data must be bivariate and the two sets must have similar variances
- The relationship between the two quantity variables must be linear

Global case study

Suppose that we wish to establish whether there is any correlation between the annual salaries of the respondents in the 'holiday survey' and the number of holidays taken per year. The analysis software that you are using will automatically calculate a PPMCC for both pairs of variables for you and it will provide the results in the form of a **correlation matrix** such as the one shown below.

	Annual salary	*Number of holidays last year*
Annual salary	1.0000	0.7592*
Number of holidays last year	0.7592*	1.0000

*SIGNIFICANCE<0.01

Figure 32 Correlation matrix

7.2.3 Evaluating results

Note that the values in the correlation matrix are correlation coefficients. Therefore the correlation between annual salaries and the number of holidays taken last year is 0.7592. This value indicates that there is a strong positive relationship between the annual salaries of respondents and the number of holidays that they took last year.

If the data were collected from a sample (as is the case here) then the software program should also give an indication of the probability that the calculated PPMCC occurred by chance. This is the SIGNIFICANCE as shown in the table above and this indicates that the probability the PPMCC of 0.7592 occurred by chance is less than 1%. Such statistics can be interpreted as meaning that the correlation between salaries and holidays is therefore highly significant.

7.2.4 Caveats about causation

As mentioned earlier, care should be exercised in drawing conclusions about correlation. You need to be sure about whether a cause and effect relationship exists or not.

- If two variables are well correlated, either positively or negatively, this may be due to pure chance or there may be a reason for it. The larger the number of pairs of data collected, the less likely it is that the correlation is due to chance, though that possibility should never be ignored entirely.

- If there is a reason, it may not be causal. For example, a person's monthly net income is well correlated with monthly deposit into a person's bank account. The logical (rather than causal) reason is that for most people, the one equals the other.

- Even if there is a causal explanation for a correlation, it does not follow that variation in the value of one variable causes variations in the value of the other. For example, sales of raincoats and of umbrellas might be well correlated, not because of a direct causal link, but because the weather influences both variables.

  Having said this, it is of course possible that where two variables are correlated, there is a direct causal link to be found.

7.3 Coefficient of determination

Unless the correlation coefficient is exactly or very nearly $+1$, -1 or 0, its meaning is a little unclear. For example, if the correlation coefficient for two variables is $+0.7592$ (as in Figure 32 above), this tells us that the variables are **positively** (but **not perfectly**) correlated, and that the correlation is **significant**. A more meaningful analysis is available from the **square of the correlation coefficient**, r^2, which is called the **coefficient of determination** (sometimes known as the **regression coefficient**).

The **coefficient of determination, r^2,** measures the proportion of the total variation in the value of one variable that can be explained by variations in the value of the other variable. It can take on any value between 0 and $+1$.

Global case study

Let us return to our example above in which we investigated the correlation between annual salaries and the number of holidays taken last year. If $r = 0.7592$, then $r^2 = 0.576$. This means that approximately 58% of the variations in the number of holidays taken by respondents last year **can be explained** by the annual salaries of those respondents and 42% of variations are probably due to other factors.

7.4 Regression equations

As we have seen, the correlation coefficient measures the **degree of correlation** between two variables, but it does not tell us how to **predict** values for one variable (y) given values for another variable (x). In order to do this, we need to determine the values of a and b in the **regression equation** ($y = a + bx$). Once your software analysis program has estimated this equation, it can be used to predict, for example, how many holidays someone earning £20,000 might have taken last year. You won't need to perform these calculations as your PC will do this for you. You should however be aware of which variables are **dependent** and **independent**.

The regression analysis equation has the form $y = a + bx$ where:

- x and y are related variables
- x is the independent variable (for example, annual salary)
- y is the dependent variable (for example, the number of holidays taken last year)
- a and b are constants which the computer program will calculate for you

Global case study

Suppose that the regression equation linking salaries and number of holidays has the following values for constants a and b.

$a = 1$
$b = 0.00005$

If we wished to predict how many holidays (y = independent variable) someone earning £20,000 (x = dependent variable) took last year, we would insert x = 20,000 into the analysis software program in order to predict the number of holidays taken.

Our results would reveal that if someone were earning £20,000 it would be estimated that they took two holidays last year.

(Note. You do not need to be able to calculate the result shown above but you might find it helpful to understand how regression analysis works by looking at the equation below.)

$Y = a + bx$

Where

$a = 1$
$b = 0.00005$
$x = 20,000$
$y = 1 + (0.00005 \times 20,000)$
$y = 2)$

7.5 Spearman's rank correlation coefficient

So far we have only looked at degrees of correlation between quantitative variables (ie variables which can be measured). Sometimes, however, variables are given in terms of order or rank rather than actual values (ordinal data). When this occurs, a different correlation coefficient, known as the Spearman's rank correlation coefficient, R, is calculated.

Spearman's rank correlation coefficient, R is used to measure the correlation between the order or rank of two variables (ordinal data). This coefficient can be interpreted in exactly the same way as the ordinary correlation coefficient (PPMCC). Its value can also range from -1 to $+1$.

8 Existence of trends

In the previous section, we studied the existence of relationships between variables and the strength and significance of the relationship. We now turn our attention to the relationships which can be identified between variables and time and how these can be used to forecast future values for the variables. This is part of a statistical technique known as time series analysis.

8.1 Time series analysis

A time series is simply a series of figures or values recorded over time. Time series can also be presented on a graph of a time series known as an historigram. The horizontal axis represents time and the vertical axis represents the values of the data recorded.

There are several features of a time series which can be identified.

- A trend
- Seasonal variations
- Cyclical variations
- Non-recurring random fluctuations

8.1.1 Trend

The trend is the most useful aspect of time series analysis applicable to business management research. The trend is the **underlying long-term movement** over time in the values of the data recorded.

Suppose a survey has been conducted over the past ten years that investigated the average amount spent on holidays per annum by 1,000 respondents living in London. It might be informative to establish what the trend in annual holiday expenditure was over that period and to use this information to predict the estimated annual holiday expenditure over the next year.

The results could be shown on a historigram, such as the one shown in Figure 33 below.

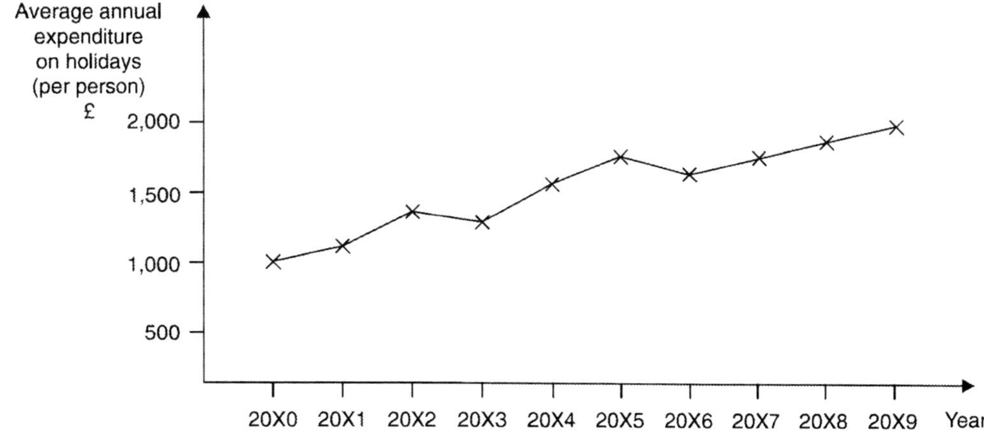

Figure 33 Historigram

The time series graph above shows that the trend is generally increasing over time, and the average annual expenditure on holidays is between £1,000 and £2,000 per person.

Here is an example of a time series which juxtaposes two so-called 'feel good factors' over a period of seven years.

Feel good factor	1998	1999	2000	2001	2002	2003	2004	Change since 1998
The nation's moral values	42	36	39	40	44	47	55	+13
The nation's economy	64	68	68	47	39	43	43	–21

The sample is drawn from a larger list of feel good factors shown in a *Harris Poll* produced on 10 November 2004, based on a survey of 1,106 US adults. The overall changes shown above have been based on the year 1998 when the overall Feelgood Index peaked at its highest point of 75.

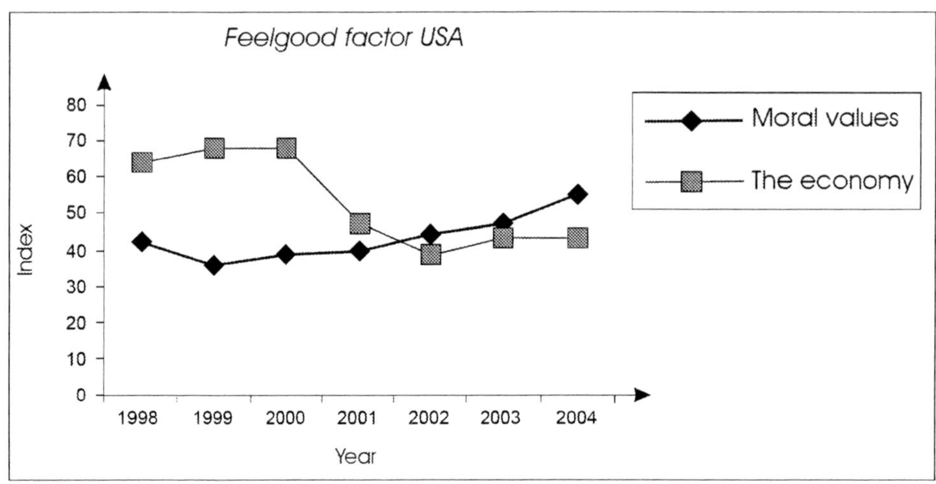

Figure 34 Moving trends

8.1.2 Moving averages

In time series analysis, the trend is found by a method known as **moving averages**. Moving averages attempts to remove seasonal and cyclical variations in order to isolate the trend. A moving average is simply an **average of several time periods**. Moving averages are relatively straightforward to calculate, and you will be relieved to hear that your analysis software will compute this effortlessly for you!

One of the most important uses of the **trend** in statistics is to use it in order to **forecast future values** by extending the trend line forward into future periods.

8.2 Index numbers

As part of your research, you might wish to compare trends between variables that are not measured in the same **units** or **magnitudes**.

Index numbers provide a standardised way of **comparing the values over time** of prices, wages, volumes and so on. They are used extensively in business, government and commerce.

Global case study

- In most countries there is a retail or consumer price index which tracks the movement of prices for a typical basket of consumer products over time (usually a monthly basis).

- A **house price index** provides an indication of the price of houses when compared to a **base period** or **base year** (which in Britain is 1983). This means that the index number for 1983 is 100 because the base period always has an index of 100.

- Producer price index which tracks so-called inflation at 'factory gate'.

- Stock exchange share prices, such as:

    - Hang Seng (Hong Kong)
    - Financial Times 100 (Britain)
    - Nikkei 225 (Japan)
    - CAC 40 (France)
    - Hermes Financial (Egypt)
    - Shanghai Composite (China)
    - Bovespa (Brazil)

If a house cost £50,000 in 1983 when the UK house price index was 100, it would be estimated to cost £260,300 in 2004 when the house price index was 520.6. The following calculation demonstrates how this figure is arrived at.

Index in base year   =   100
Index in 2004   =   520.6
Price in base year   =   £50,000

$$\therefore \text{Price in 2004} = \frac{\text{price in base year} \times \text{index in 2004}}{\text{index in base year}}$$

$$= £260,300$$

We could say that the house price index has risen 420.6 points (520.6 – 100) between 1983 and 2004 and that this represents an increase of 420.6%.

If we wished to compare the costs of items of expenditure of the average household (the RPI) with the average cost of a house, we could do this by comparing the index values for each variable. In recent years, these indices have made it clear that the cost of buying a house in Britain has risen at a much greater rate than the cost of the average family's weekly shopping. As with all of the techniques that we have covered in this chapter, you must be able to interpret the results obtained. Do make use of the analytical tools available or software to help you prepare diagrams and compute key statistics.

1 Distinguish between types of quantitative data

- Quantitative data can be classified as nominal, ordinal, interval or ratio.

- The level of measurement will determine the statistical methods used.

2 Evaluate the use of alternative data collection tools in your research project

- Questionnaires are usually divide into two categories.

  – Self administered

  – Interviewer administered.

- The problem of non-respondents in administered questionnaires must be borne in mind.

- Questionnaires should be pilot tested; otherwise the risk of failure is high.

- When designing questionnaires the layout, questions asked and approach to respondents need careful consideration.

3 Prepare your quantitative research data for analysis

- After data has been collected it must be prepared for analysis. This may involve editing, cleaning, matrix insertion, coding and weighting of cases.

4 Identify appropriate graphical techniques for preliminary data review

- Data may be presented graphically with the help of tools, such as frequency distributions, bar charts, histograms, line graphs and pie charts.

5 Measure central tendency and dispersion

- Measures of central tendency – commonly known as averages.

- Measures of dispersion – assess how dispersed around the average the variables are.

6 Identify appropriate statistical techniques to apply within your data analysis

- There are a number of stages associated:

  Step 1   Specify the null hypothesis

  Step 2   Specify the alternative hypothesis

  Step 3   Establish the level of statistical significance

  Step 4   Select the appropriate statistical test

  Step 5   Test for statistical significance

- Statistical tests chosen are dependent on the number of groups, level of measurement and whether the variables are related or independent.

7 Establish the existence of relationships in data

- Correlation and regression establishes whether relationships exist between data.

8 Establish the existence of trends

- Time series analysis can be used to demonstrate the behaviour of a variable over time.

1   This will depend on your own research.

2   Analysis of variance. There are four independent groups and the data has interval/ratio properties. Table C would provide the answer. Test some other scenarios for yourself.

Bell, J., (2005). *Doing Your Research Project: A Guide for First Time Researchers in Education, Health and Social Science.* 4th edition. Buckingham: Open University Press.

Bryman, A. and Bell, E., (2007). *Business Research Methods.* 2nd edition. Oxford: Oxford University Press.

Collis, J. and Hussey, R., (2003). *Business Research: A Practical Guide for Undergraduate and Postgraduate Students.* 2nd edition. Basingstoke: Macmillan Polgrave.

Denscombe, M., (2003). *The Good Research Guide for Small-Scale Social Research Projects.* 2nd edition. Buckingham: Open University Press.

Diamantopoulous, A. and Schlegelmilch, B.B., (2000). *Taking the Fear Out Of Data Analysis.* London: Thomson.

Dillon, W., Madden, T. & Firtle, N., (1994). *Marketing Research in a Marketing Environment*, 3rd Edition. Illinois: Irwin .

Easterby-Smith, M., Thorpe, R. and Lowe, A., (2002). *Management Research: An Introduction.* 2nd edition. London: Sage.

Gill, J. and Johnson, P., (2002). *Research Methods for Managers.* 3rd edition. London: Sage.

Hammerton, M., (1975). *Statistics for the Human Sciences.* London: Longmans.

Kazmier, L.J., (2003). *Theory and Problems of Business Statistics.* 4th edition. New York: McGraw-Hill.

Phelan, P. and Reynolds, P , (1996). *Argument and Evidence; Critical Analysis for the Social Sciences.* London: Routledge.

Saunders, M., Lewis, P. and Thornhill, A., (2009). *Research Methods for Business Students.* 5th edition. Harlow: Pearson Education Limited.

Sekaran, U., (2002). *Research Methods for Business: A Skills Building Approach.* 4th edition. New York: Wiley.

Wallace, R.S.O. and Mellor, C.J., (1988). 'Non-Response Bias in Mail Accounting Surveys: A Pedagogical Note.' *British Accounting Review* 20.

Weiss, N.A., (2002). *Elementary Statistics.* 3rd edition. Reading, Massachusetts: Addison-Wesley.

Zuaniecki, F., (1934). *The Method of Sociology.* New York: Farrar and Rinehart.

Zikmund, W.G., (2002). *Business Research Methods.* 4th edition. South-Western College publishers.

Research Methods for Global Marketing Practice

Index

Research Methods for Global Marketing Practice

Insufficient data, 105
Integrity, 110
Interactive approach, 69
Inter-library loan, 89
Internal logic, 18
Internet, 66, 69, 71
Interview, 121
Interviewer bias, 124
Intrusive information, 113
Iterative, 30
Iterative approaches, 31

Jargon, 66

Journals, 55, 56, 66, 92
Judgemental sampling, 101

Key references, 71

Key words, 68, 69, 72
Knowledge, 9

Large samples, 17

Leadership styles and project management, 46
Learning approach, 31
Leisure Tracking Survey, 91
Libraries, 71
Library, 56, 71, 89, 92
Linked bonus scheme, 5
Literature review, 35, 63, 65, 67, 74
Literature search, 63, 67, 69, 74
Literature survey, 62, 63, 73
Logical relationships, 2
Logistics, 34
Longitudinal studies, 26
Loyalty card scheme, 2

Management accounts, 65

Management by exception, 46
Management consultancy, 7
Management consulting reports, 66
Management practice, 6
Management problems, 6
Management problem-solving, 5
Management research, 2, 6, 7, 11, 21, 30, 108
Managing change, 7, 20
Market research, 87, 89
Market research projects, 65
Mash up, 133
Mathematical model, 64
Mean, 190
Median, 191
Methodology, 59, 114

Middle range theories, 13
Milestones, 34
Mind maps, 139
Mission statements, 11, 65, 66
Mixed methods approach, 87
Mode, 190
Models, 13, 63
Moderator, 130
Monographs, 66
Montevideo, 68
Motivation, 57
Multi-concept, 68
Multi-disciplinary, 6, 68
Multiple-case study, 24
Multiple-methods, 21
Multi-stage process, 31
Multivariate, 197
Multivariate data analysis, 197

Narrow objectives, 77

Negotiation, 42
Negotiation process, 43
Negotiation skills, 42
Nonmaleficience, 108
Non-random sampling, 101
Non-responses, 106, 159
Notebook, 57
Null hypothesis, 193, 200, 201, 204

Objectivity, 11, 113, 115

Observed phenomena, 12
Observer effect, 22
Official statistics, 93
Omission, 105
Open coding, 142
Operationalisation, 15
Organic structure, 87
Organisation, 34
Organisational culture, 11

Pamphlets, 65

Paradigm, 8, 11
Paradigm wars, 11
Paradoxes, 56, 74
Parsimony, 9
Partial correlation, 189
Participant observation, 111
Pattern-matching, 140
Pearson's product moment correlation
    coefficient, 210
Personal bias, 2, 115
Personal contacts, 84

 Research Methods for Global Marketing Practice